THIRD EDITION

NELSON

VISUAL COMMUNICATION DESIGN

VCE UNITS 1-4

WORKBOOK

KRISTEN GUTHRIE

Nelson Visual Communication Design VCE Units 1–4 Workbook
3rd Edition
Kristen Guthrie

Publishing editor: Debbi Barnes
Project editor: Kathryn Coulehan
Editor: Katharine Day
Proofreader: Bree DeRoche
Permissions researcher: Debbie Gallagher
Text design: Aisling Gallagher
Cover design: Kristen Guthrie
Production controller: Karen Young
Typeset by: MPS Limited

Any URLs contained in this publication were checked for currency during the production process. Note, however, that the publisher cannot vouch for the ongoing currency of URLs.

Acknowledgements
The publisher would like to credit and acknowledge the following source for the photograph on pages 50 and 205 (top left): Ball Chair, design Eero Aarnio, photo Harri Kosonen, Studio Sempre.

© 2017 Kristen Guthrie

For product information and technology assistance,
in Australia call **1300 790 853**;
in New Zealand call **0800 449 725**

For permission to use material from this text or product, please email
aust.permissions@cengage.com

ISBN 978 0 17 040179 1

Cengage Learning Australia
Level 7, 80 Dorcas Street
South Melbourne, Victoria Australia 3205

Cengage Learning New Zealand
Unit 4B Rosedale Office Park
331 Rosedale Road, Albany, North Shore 0632, NZ

For learning solutions, visit **cengage.com.au**

Printed in China by 1010 Printing International Limited.
5 6 7 8 25 24 23 22

CONTENTS

A

Thinking

B

Drawing & designing visual communications

ISBN 9780170401791

Writing about visual communications

Applying the design process

ISBN 9780170401791

INTRODUCTION

This workbook is designed as a companion to the *Nelson Visual Communication Design VCE Units 1–4* student book (Fourth edition). This book features tasks that are specifically designed to help develop skills in essential areas of the VCE Visual Communication Design coursework.

It is structured to build skills in fundamental course concepts including:

Drawing methods • Observation • Visualisation • Presentation	Two-dimensional	• Packaging nets • Orthogonal drawing	
	Three-dimensional	Paraline drawing	• Isometric • Planometric
		Perspective drawing	• One-point perspective • Two-point perspective
	Rendering		
Design elements	Colour		
	Form		
	Line		
	Point		
	Shape		
	Texture		
	Tone		
	Type		
Design principles	Balance		
	Contrast		
	Cropping		
	Figure–ground		
	Hierarchy		
	Pattern		
	Proportion		
	Scale		
Design thinking			
Visual communication terminology			
Visual communication analysis			
Design fields			
Writing a design brief			
Applying the design process			
Evaluation			

Content in this workbook is designed to assist in building the skills required in the *VCE Visual Communication Design Study Design 2018–2022*.

ISBN 9780170401791

HOW TO USE THIS WORKBOOK

This workbook is flexible and information contained in each section can be applied at any stage of Units 1 to 4 for skill building or revision; it is arranged thematically. You can choose to complete some exercises manually and some using design and drawing software; it is up to you.

The table below illustrates how the Nelson Visual Communication Design workbook and student book work together to cover all aspects of the *VCE Visual Communication Design Study Design 2018–2022*.

Workbook chapter	Student book chapter(s)
1 Design thinking tools	9 Design thinking
2 Applying design elements and design principles	5 Design elements and design principles
3 Purposes of visual communications	4 Purposes of visual communications
4 Contexts of visual communcation	5 Design elements and design principles 12 Design in a wider context
5 Audiences	10 Visual language and analysis
6 Two-dimensional drawing	1 2D and 3D drawing
7 Three-dimensional drawing	1 2D and 3D drawing
8 Rendering textures and materials	2 Observational drawing and rendering
9 Typography and layout	6 Typography and layout
10 Analysis of visual communications	10 Visual language and analysis
11 Design fields	11 Design fields
12 Intellectual property	13 Legal and ethical issues in design
13 Writing the design brief	8 Design process 10 Visual language and analysis
14 Research	8 Design process
15 Generation, development and refinement	8 Design process 9 Design thinking
16 Evaluation and production	8 Design process

ISBN 9780170401791

To assist you in using this book the following symbols are placed beside each task.

 This symbol indicates that you should complete the task in the workbook.

 This symbol indicates that you should complete the task on a separate piece of paper or in your visual diary or sketchbook.

 This symbol indicates extra 'Takeaway' tasks that you can do at home.

All technical drawing tasks follow the VCAA *Technical Drawing Specifications*.

Grids have been provided in Appendix I at the back of this book to assist you in applying both orthogonal and paraline drawing methods.

Resources that will help you make the most of this workbook are:

- *Nelson Visual Communication Design VCE Units 1–4*
- *VCE Visual Communication Design Study Design 2018–2022*
- VCAA *Technical Drawing Specifications*.

AUTHOR ACKNOWLEDGEMENTS

Many freehand illustrations throughout the workbook were created by Mark Wilken of Studio Workshops www.studioworkshops.com.au

This book was made possible thanks to the assistance of some exceptional people; in particular Mark Wilken and my very talented Woodleigh School VCE Visual Communication Design students.

Many thanks to my supportive colleagues at Woodleigh School. A big thanks to the Nelson Cengage team.

ISBN 9780170401791

PART A

Chapter 1
DESIGN THINKING TOOLS

To initiate and develop creative design ideas, a range of strategies can be applied. From 'blue-sky' thinking techniques where the sky is the limit to tools that assist in identifying the most appropriate design concept; design thinking tools offer structures to stimulate ideas and seek solutions. You will find that Chapter 9 of *Nelson Visual Communication Design VCE Units 1–4* will help you in completing this chapter.

1.1 CONCEPT MAPS

Concept maps are a popular tool for the brainstorming and organisation of initial design ideas. The map may be as simple or complex as you like but it should be rich with information to suggest design directions.

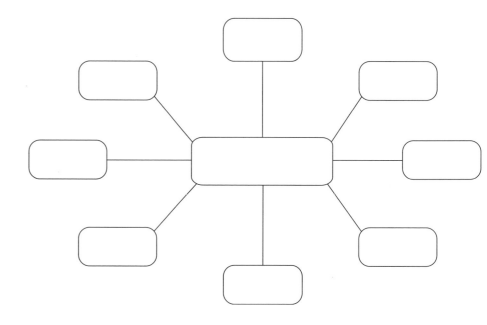

ISBN 9780170401791

1.2 WORD LISTS

Making a list of words and phrases that are related to the design problem is an effective strategy. Use the template below to start your own list.

Where will the design be located/ used? List all the possible locations and uses.	Who uses or sees the design? List every person or group that may come into contact with the design.	What does the design need to do/achieve? List all the possible functions that the design could achieve.	What features might the design include? Be creative in listing every possible feature, practical or not. There are no right or wrong answers.	How will the design be distinctive? What could be done to the design to ensure it stands out?

1.3 RANDOM WORDS

Photocopy the card template on the following pages to create your own brainstorming toolkit. When you are stuck for ideas, randomly pick a card to initiate a design direction. Use the words on the next page or create your own using the blank template on page 5.

ISBN 9780170401791

FLIGHT	ECO	DOUBLE	TRANSFORM
LIGHT & DARK	EXCHANGE	CONNECTION	ABSTRACT
HIDDEN	PATTERNED	POSITION	SILENT
HOLLOW	LADDER	WEATHER	REFLECTION
PRISM	COMPLEX	BIND	GRID
STICKY	FLAT	WATER	CYCLE
BLENDED	SEAT	CONTAIN	CLAW

ISBN 9780170401791

ISBN 9780170401791

1.4 RUSSIAN DOLL

Russian doll is a technique that 'splits' parts of a design problem into different attributes, which can lead to new and unexpected ideas. Use the template to create your own breakdown of possible design directions using this technique.

DESIGN REQUIREMENTS	ATTRIBUTES X 2	ATTRIBUTES X 4

ISBN 9780170401791

1.5 IDEA BOX

The idea box assists in making visual and functional connections at the early stages of the design process. Noting the characteristics of the design task at the top, list possible variations in each column. Then make connections between each column to stimulate new design combinations.

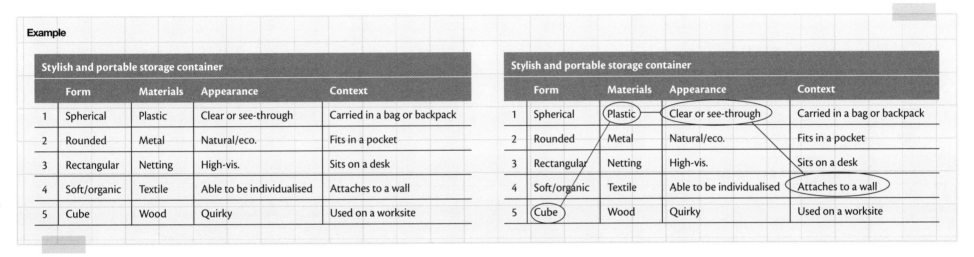

Example

Stylish and portable storage container			
Form	**Materials**	**Appearance**	**Context**
1 Spherical	Plastic	Clear or see-through	Carried in a bag or backpack
2 Rounded	Metal	Natural/eco.	Fits in a pocket
3 Rectangular	Netting	High-vis.	Sits on a desk
4 Soft/organic	Textile	Able to be individualised	Attaches to a wall
5 Cube	Wood	Quirky	Used on a worksite

Stylish and portable storage container			
Form	**Materials**	**Appearance**	**Context**
1 Spherical	Plastic	Clear or see-through	Carried in a bag or backpack
2 Rounded	Metal	Natural/eco.	Fits in a pocket
3 Rectangular	Netting	High-vis.	Sits on a desk
4 Soft/organic	Textile	Able to be individualised	Attaches to a wall
5 Cube	Wood	Quirky	Used on a worksite

Idea box for the design of a stylish, portable storage container that accommodates cables, chargers and other connections for mobile devices and laptops.

ISBN 9780170401791

Create your own idea box.

	Form	Materials	Appearance	Context
1				
2				
3				
4				
5				

ISBN 9780170401791

1.6 SCAMPER

SCAMPER is a powerful tool that can be used in the development of a design idea. It can bring about innovative new directions during the design process as it asks challenging questions about a concept.

SCAMPER elements	Key questions to ask	What are the possible results in your design?
Substitute	What if I swap this for that and see what happens?	
	Who else could find this appealing or useful?	
	What other materials or design factors could I use instead?	
	What happens if I substitute the shape, texture, form or colour?	
Combine	What elements or principles of design can be combined?	
	What graphical representations could be combined?	
Adapt or Add	What part of the concept can I change?	
	What if I were to use parts of other design elements or principles?	
	What if I reuse aspects of my design in other ways or other places?	
Modify or Magnify or Minimise	What happens if part of the concept is expanded, exaggerated, minimised or changed?	
	What is the effect of altering proportions and relationships in the design?	
Put to another use	What other function or use can my concept be applied to?	
	Can another design feature from another product be used in my idea?	
Eliminate or Erase	What can be removed from my concept?	
	What can be understated or streamlined?	
	What happens to the design if parts are taken away?	
Reverse or Rearrange	What is the opposite of what I am currently doing?	
	What if I did it the other way around?	
	What if I reverse the elements or the way it is used?	
	What happens if I mix up the design?	

ISBN 9780170401791

1.7 PLUS, MINUS, INTERESTING

When deciding on a design direction, it can be challenging to separate ideas and to leave some behind. Using the Plus, Minus, Interesting organiser, arrange the positive and negative aspects of a design idea into categories. Use the 'interesting' section to identify what might be changed when developing this idea further.

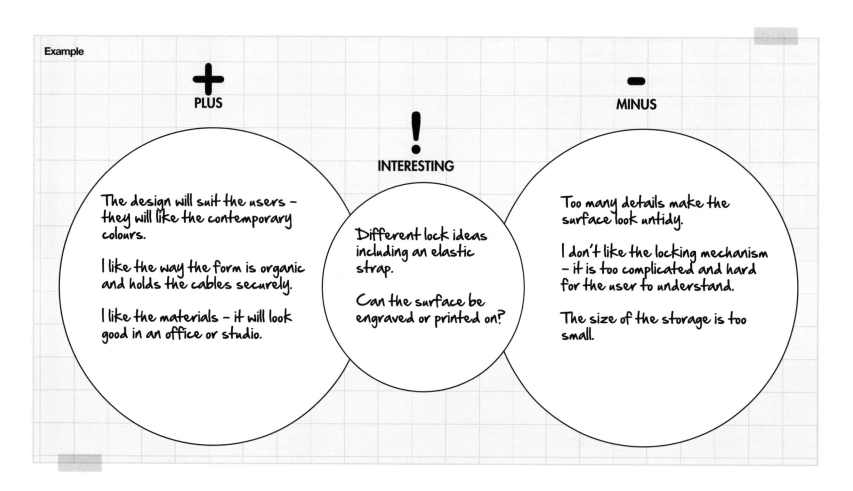

Example

+
PLUS

!
INTERESTING

−
MINUS

The design will suit the users – they will like the contemporary colours.

I like the way the form is organic and holds the cables securely.

I like the materials – it will look good in an office or studio.

Different lock ideas including an elastic strap.

Can the surface be engraved or printed on?

Too many details make the surface look untidy.

I don't like the locking mechanism – it is too complicated and hard for the user to understand.

The size of the storage is too small.

ISBN 9780170401791

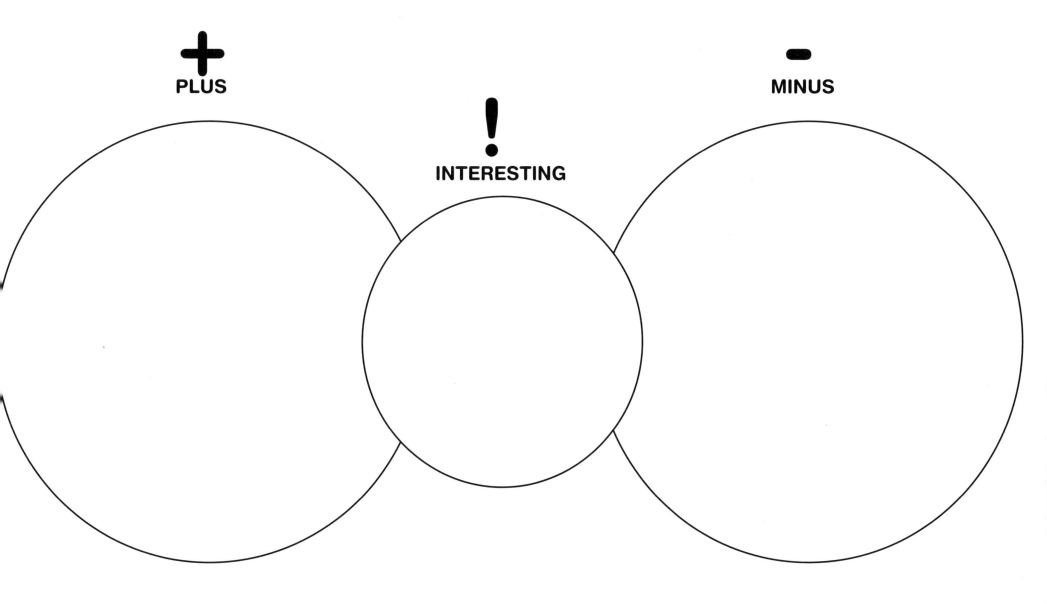

ISBN 9780170401791

1.8 POOCH

The POOCH model assists in critical thinking and decision making. POOCH can be used to help choose between options during the design process.

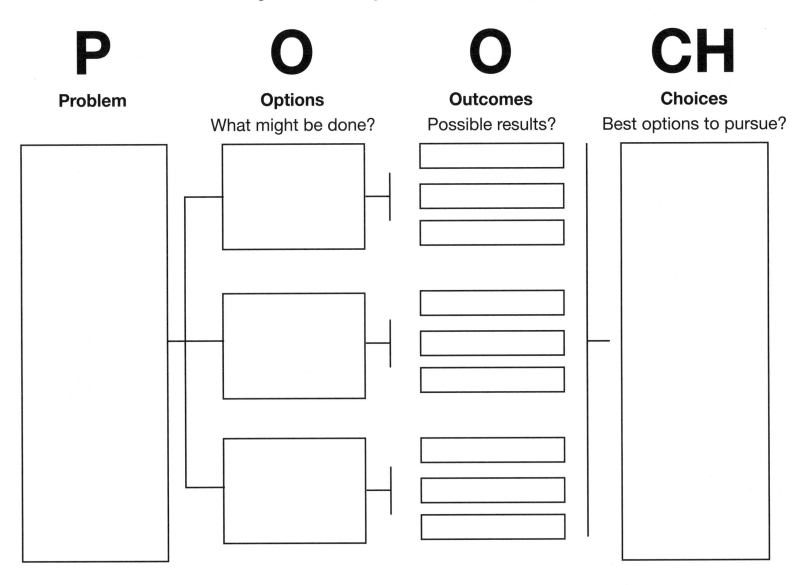

P

Problem

O

Options
What might be done?

O

Outcomes
Possible results?

CH

Choices
Best options to pursue?

ISBN 9780170401791

1.9 SWOT ANALYSIS

A SWOT analysis is a framework for analysing the strengths and weaknesses of a design, and the opportunities and threats (concerns) that it raises. It helps to focus on strengths, address concerns and take advantage of opportunities that are identified.

ISBN 9780170401791

Chapter 2

APPLYING DESIGN ELEMENTS & DESIGN PRINCIPLES

Design elements and design principles are important in the effective design of visual communication. They are the essential components of visual compositions; it is vital that you have a strong understanding of their meaning and skill in their application.

Chapter 5 of *Nelson Visual Communication Design VCE Units 1–4* will help you with this chapter.

Design elements	Design principles
Colour	Balance
Form	Contrast
Type	Cropping
Line	Figure–ground
Point	Hierarchy
Shape	Pattern
Texture	Proportion
Tone	Scale

2.1 DESIGN ELEMENTS

Design elements are the building blocks that we use to construct a composition; in fact, they are the fundamental components of a composition.

The exercises in this section are designed to help you become familiar with some of the characteristics of the design elements of VCE Visual Communication Design.

Colour

 **TASKS**

1   Complete the colour wheel and colour swatches. Name them appropriately.

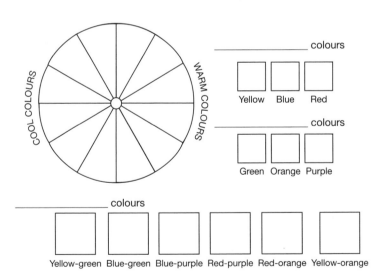

_____ colours
Yellow   Blue   Red

_____ colours
Green   Orange   Purple

_____ colours
Yellow-green   Blue-green   Blue-purple   Red-purple   Red-orange   Yellow-orange

ISBN 9780170401791

2 Use a digital camera to take images of colour and use appropriate computer software to create a seasonal colour palette. Annotate your imagery by describing how the colour palette might be applied to different areas of design; for example, fashion, architecture or graphic design.

3 Add colour to the pictured interior illustrations. One illustration should use analogous colours while the other should use complementary/contrasting colours. Describe below the impact of the different colour selections on the appearance of the interior.

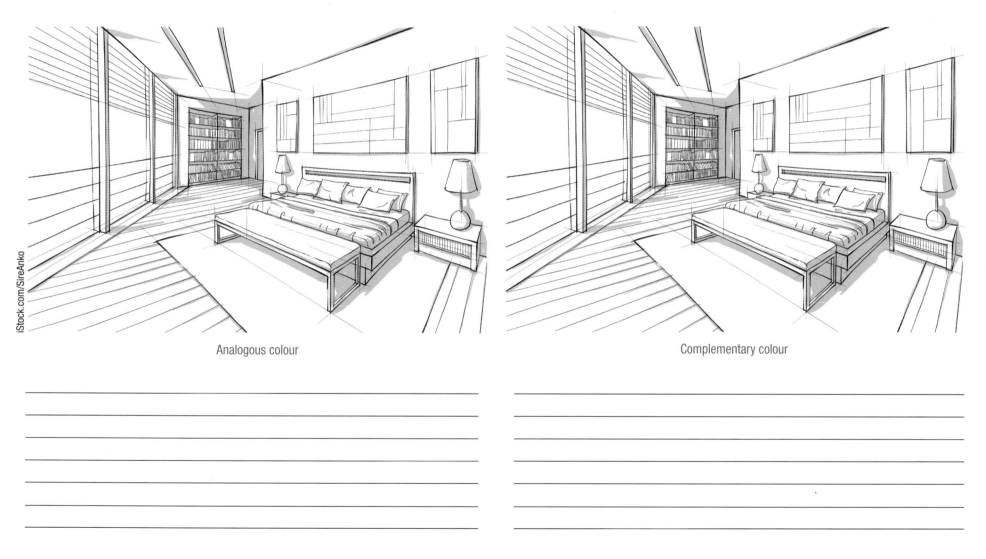

iStock.com/SireAnko

Analogous colour   Complementary colour

_____   _____

_____   _____

_____   _____

_____   _____

_____   _____

_____   _____

_____   _____

ISBN 9780170401791

Form

Generally the term form is used to describe objects that are three-dimensional in nature.

 TASKS

1   Using tone, colour, texture and the given light source, transform these objects into rendered three-dimensional forms.

Light source

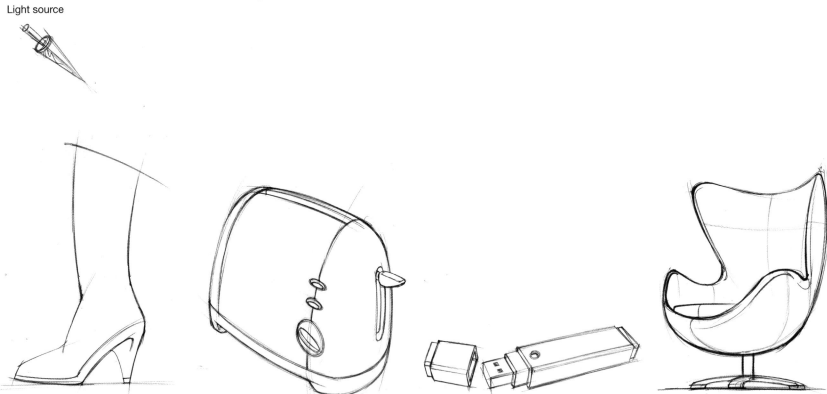

ISBN 9780170401791

2 Look at the packaging template (a two-dimensional plan development drawing). Visualise the template as a three-dimensional object and draw it in its constructed form. You may add decoration as you like. Alternatively, you may wish to construct the packaging from paper or card.

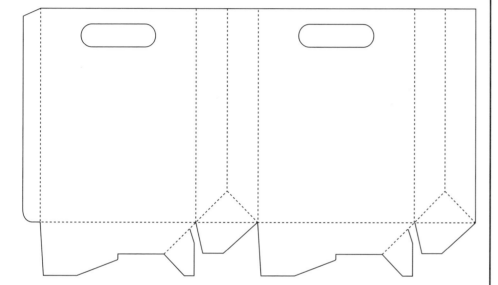

Type

 TASKS

1 Use your ruler to indicate the x-height and baseline of the type. Then indicate the following features:

- upper-case type
- ascenders
- descenders
- leading
- spacing
- serifs
- lower-case type.

Typography & Design.

2 Match the illustrated types to the appropriate typographic style. Write your answers on the lines provided.

- serif
- sans serif
- decorative
- handwritten
- script
- slab serif.

1. Typography   _____
2. Typography   _____
3. **Typography**   _____
4. Typography   _____
5. *Typography*   _____
6. **Typography**   _____

ISBN 9780170401791

Line

Line is used widely in visual communication for both technical and creative applications.

 TASKS

1 The poster below lacks visual connection between the title at the top and the illustration at the bottom. Using line only, creatively connect the two separate aspects of the poster to create an effective composition. Maintain a balanced composition.

2 Create a pattern using line as the dominant element. Apply your pattern to two different objects (use the objects pictured or create your own).

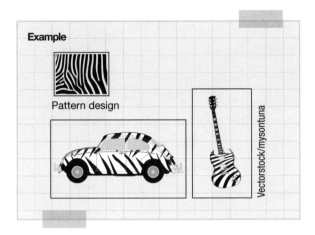

Example

Pattern design

Vectorstock/mysontuna

Pattern design

ISBN 9780170401791

Point

Point is often used in diagrams and on maps to indicate key information.

 TASKS

1   On the human skeleton, indicate these
    body parts using point:

    - ankle
    - wrist
    - spine
    - rib cage
    - skull
    - pelvis
    - knee
    - elbow.

    Annotate your use of point
    and explain how it effectively
    communicates the required
    information to the viewer.

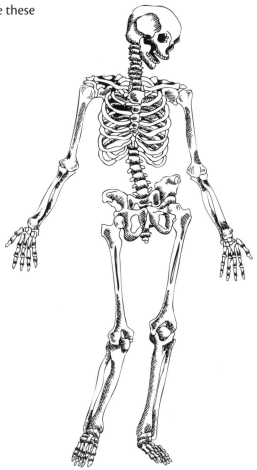

2   A wildlife rescue centre offers tours of feeding times. A shuttle service will take
    visitors to the different animal zones at different times of the day and evening
    to watch feeding. On the map use point to create a system that assists visitors
    in navigating to each feeding event. There are four proposed tours on offer.
    Also, consider the use of colour to ensure clear communication.

    - Morning tour: koalas, birds of prey and emus.
    - Afternoon tour: kangaroos and wallabies, wombats and platypuses.
    - Evening tour: quolls, bilbies and bandicoots, and possums.
    - VIP private donor tour: animal nursery and all animal areas.

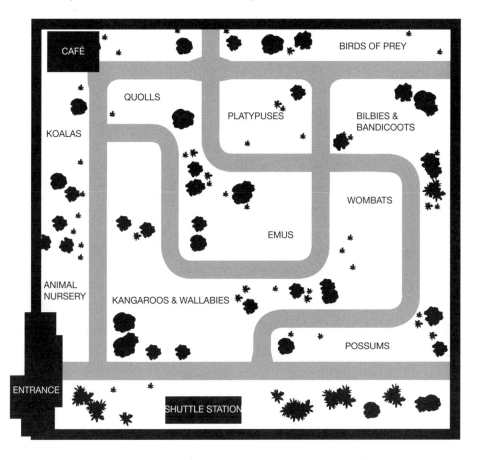

ISBN 9780170401791

Shape

Shape usually refers to abstract or representational shapes that are two-dimensional.

 TASKS

1   Caution signs are a common sight on our roads. Create three new signs that communicate the need for caution and awareness of slightly unusual hazards. Use only symbols (see example).

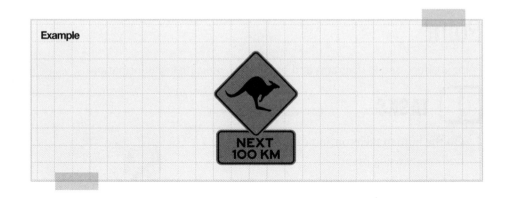

Example

Dancing Zone Ahead   Cane Toad Crossing   Hang Gliders Above

ISBN 9780170401791

2   Identify the meanings of the following symbols.

_____   _____   _____   _____   _____

_____   _____   _____

3   Pictured on the right is a simple step-by-step diagram of how to tie a necktie. Using a similar graphic style, create a diagram in the space below that explains how to construct a paper plane. Use only shape and do not include text.

HOW TO TIE A TIE
FOUR-IN-HAND KNOT

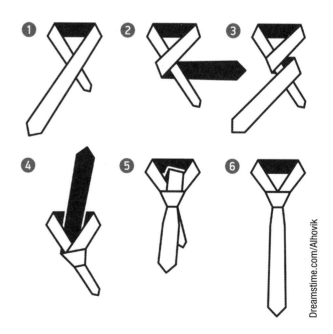

Dreamstime.com/Alhovik

Texture

 TASKS

1 Add the following textures to the image of home furnishings:

- chairs: leather with chrome legs
- table: wood with chrome legs and rubber feet
- light: plastic shade with chrome support and solid wood base
- bottle and glasses: glass.

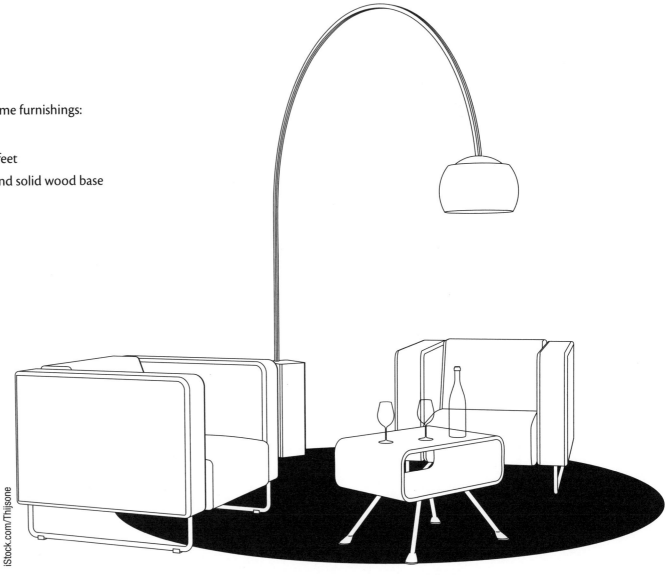

iStock.com/Thijsone

ISBN 9780170401791

2 Texture can be created with line and shape. Using some of the illustrated textures in the example, create a graphic design for the front of a T-shirt. Use texture only to create an effective and appealing composition.

Example

ISBN 9780170401791

Tone

 TASKS

1 You are required to render the illustrations of sneakers.

a Render this illustration with only flat colour. Do not indicate the light source or shadow.

b Render this illustration with tone and colour using the light source to guide your application. You may also indicate the cast shadow.

Light source

c In the space provided, explain the key differences between the two rendered images and suggest a context where each would be appropriate.

_____

_____

_____

_____

_____

_____

_____

_____

_____

2 You will need a torch (flashlight) and a mostly white object, such as an egg, garlic or a parsnip.

Using white paper as a background, position the torchlight so that it illuminates the object and casts obvious shadows. On thick paper stock (150gsm or higher) and using only a 4B or 6B pencil, render the form of the object. Use tone _only_ and _do not outline_ the shape of the object. See example below:

Thomas Rennie

ISBN 9780170401791

TAKEAWAY

1   Collect one example of visual communication for each design element. Select examples that display each of the elements in an effective way. Annotate your selections with the following information:

- Where would you locate the visual communication (include source attributions)?
- Who is the target audience of the visual communication?
- Give a description of the appearance of the design element, suggesting a reason for its application to the visual communication.

2.2 DESIGN PRINCIPLES

If design elements are the building blocks, design principles are the construction techniques we use to construct a composition. Design principles are the considerations and design 'rules' applied to a composition.

Balance

TASKS

1   Using the images provided for you, create two alternative compositions that clearly depict symmetrical balance and asymmetrical balance. You may photocopy the images to adjust the scale or repeat elements. You may add other design elements as required to create an imaginative composition.

iStockphoto/Talshiar

20th CENTURY MODERN

Symmetrical balance

Asymmetrical balance

ISBN 9780170401791

2 An art gallery housed in a historic building has approval for a contemporary extension. Using a range of design elements, create an extension that provides balance to the original building. Strive to create a cohesive, complementary and contemporary architectural design.

ISBN 9780170401791

Contrast

TASKS

1 Using two circles or squares (see example), select ideas from the list to create a composition that explains contrast:

- stable
- smooth
- quiet
- hard
- dry

- light
- increasing
- transparent
- mild
- concave

- unstable
- rough
- loud
- soft
- wet

- heavy
- decreasing
- opaque
- bold
- convex.

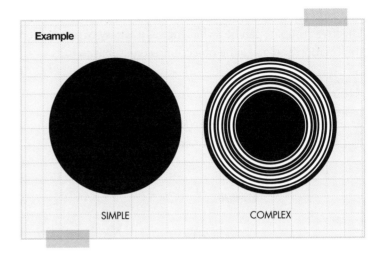

Example

SIMPLE   COMPLEX

2 Apply contrast as indicated to these three compositions.

Create contrast using line

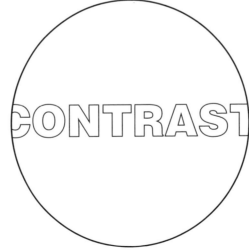

Create contrast using colour

Create contrast using tone

ISBN 9780170401791

Cropping

Effective cropping can be a challenge as it is easy to remove too much visual information and make the object unrecognisable.

TASKS

1 Crop the images (below) effectively and ensure that their visual meaning is retained (see example).

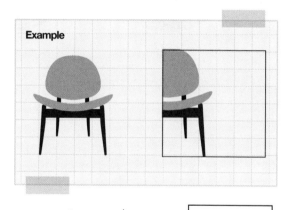

Example

1

2

3

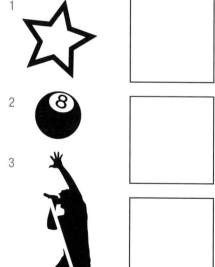

2 Create six 50 × 50 mm squares on paper. Select six words from the following list and create a series of six cropped images in your book, similar in style to the examples. You may use black and white or colour. You may choose any media.

Select six words from this list to create your cropped images:

- aeroplane
- shoe
- banana
- koala
- sunshine
- beach
- palm tree
- eagle
- fish
- boat
- key
- pig
- spaghetti
- storm
- truck.

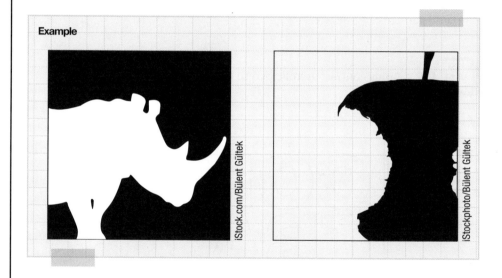

Example

iStock.com/Bülent Gültek

iStockphoto/Bülent Gültek

ISBN 9780170401791

Figure–ground

TASKS

1   Create two compositions from the images provided. The first should use figure–ground to emphasise the figure. The second should utilise figure–ground to de-emphasise the figure. You may use the photocopier to enlarge or reduce elements and you may add any other elements or media that assist in creating effective compositions.

2   Observe the illustration and describe the relationship between the figure and ground.

_____

_____

_____

_____

_____

_____

Use the figure–ground relationship to de-emphasise the television.

Use the figure–ground relationship to emphasise the television.

iStockphoto/Bülent Gültek

ISBN 9780170401791

Hierarchy

 TASKS

1 Using numbers 1–4, identify the four most dominant elements in the hierarchy of the pet show poster. Annotate your answer by explaining the reasoning behind your selections.

2 Pictured is a promotional shirt with no visible *hierarchy*. Use the images provided to create two alternative hierarchies for this business; the hierarchy of each design is described beneath the template. You may photocopy the images and paste them into each composition along with other elements of your choice or draw your concepts by hand.

ISBN 9780170401791

Emphasise: 1 business name, 2 imagery

Emphasise: 1 imagery, 2 business name

ISBN 9780170401791

Pattern

 TASKS

1 Create two patterns, one using repetition and one using alternation. Apply each pattern design to the boots. You may use a medium of your choice and apply a range of elements, such as colour, line, tone and shape, in the execution of your patterns.

Pattern 1: repetition   Pattern 2: alternation

2 Create one repetitive pattern and one alternating pattern from these naturally-occurring patterns. Use line, colour, shape and tone to create your patterns.

Pattern showing repetition

Pattern showing alternation

ISBN 9780170401791

Proportion

Proportion is about relationships between visual elements. Proportion can be used to create harmonious compositions or may be altered to exaggerate and emphasise elements.

 TASKS

1 Pictured is the logo for 'A Little Pizza Heaven'. Using the logo and appropriate typography and symbols, create a simple pizza order app for use on mobile devices. Adjust the proportions of the app content and design to suit both a smart phone and tablet.

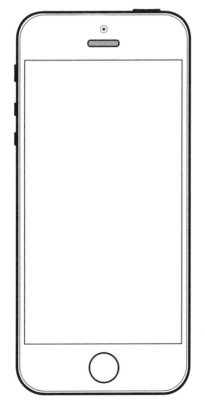

ISBN 9780170401791

2 Pictured is a floor plan of a new café. The structure of the building is complete but tables and chairs need to be added. Considering *proportion* as well as function, add chairs and tables to the floor plan while allowing for 'flow' and safe access to essential services including:

- front entrance
- rear exit
- staff access to tables, kitchen and coffee bar
- bathroom access for patrons.

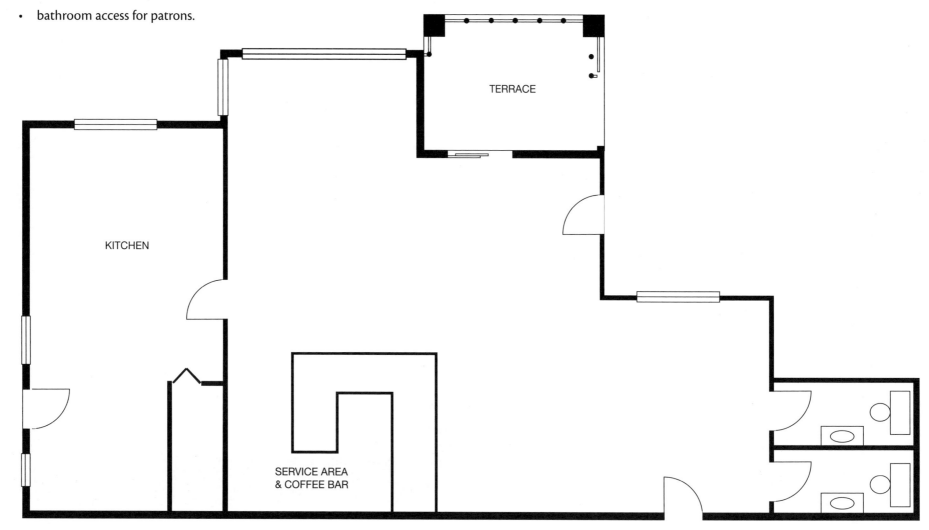

ISBN 9780170401791

Scale

Scale refers to the size of elements within a composition. Scale can establish clear visual relationships and create a compositional hierarchy.

TASKS

1   Using the images provided, create a diagram that uses scale to communicate the level of danger that each animal poses to visitors to Australia. You may use the photocopier to enlarge the images or draw them freehand.

iStockphoto/Jelena Katkova

2   The illustration of a pencil has been drawn at a scale of 1:1. Redraw the pencil using the following scales:

a   2:1

b   1:2

TAKEAWAY

1   Collect one example of visual communication for each design principle (eight in total). Select examples that display each of the principles clearly. Annotate your selections with the following information:

- where you found it (the context)

- what the visual communication is: e.g., packaging or an advertisement

- a description of the appearance of the design principle (including the use of design elements within the composition), suggesting how this principle assists in communicating visual information.

ISBN 9780170401791

2   Use the templates provided on the next two pages to create two composition series that clearly convey the meaning of all of the design elements and all of the design
    principles. Use ICT, cut paper or a preferred medium of your choice. See examples below.

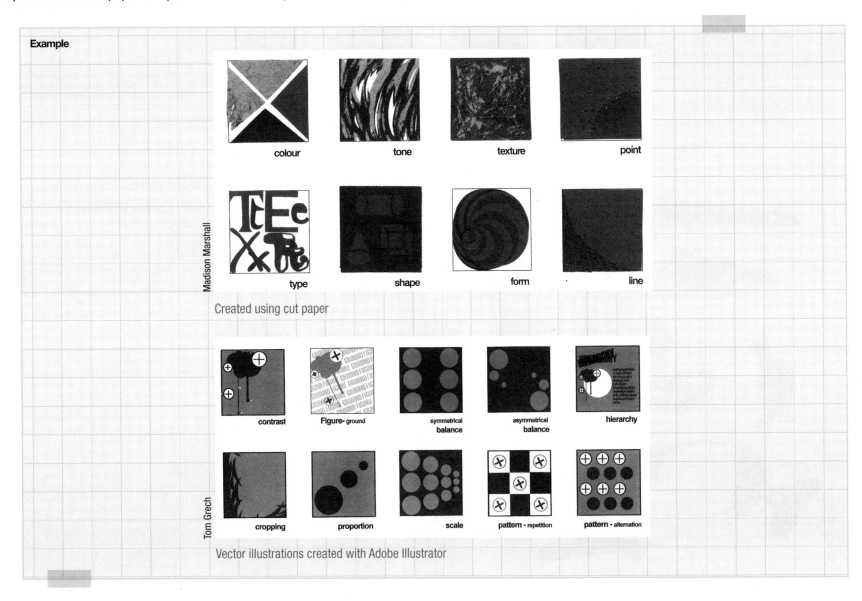

Colour

Tone

Texture

Point

Type

Shape

Form

Line

ISBN 9780170401791

Contrast

Figure-ground

Symmetrical balance

Asymmetrical balance

Hierarchy

Cropping

Proportion

Scale

Pattern (repetition)

Pattern (alternation)

ISBN 9780170401791

Chapter 3

PURPOSES OF VISUAL COMMUNICATIONS

Visual communications have one or more purposes, which have an impact on their content and appearance. The purpose may define the content of the visual communication and establish where and how it will be seen or used, who will see it/use it, and how often. Chapter 4 of *Nelson Visual Communication Design VCE Units 1–4* will help you with this chapter.

TASKS

1 Suggest the most suitable purpose for the following visual communications. Identify the purpose or purposes and explain your choice. You can find colour versions of the images in the table in Appendix III (pages 203–4).

Visual communication	Purpose(s)	Explain how the visual communication achieves it purpose or multiple purposes
Atelier Art Supplies Pty Ltd — ATELIER ART SUPPLIES	To identify	The logo for Atelier Art Supplies uses the shape of a square, placed within the `L' to suggest a painter's easel. Additional shapes that help to suggest the legs and top of the easel reinforce the nature of the business: art materials.
Emma Rickards — Design principles poster for use in the Visual Communication Design classroom		

ISBN 9780170401791

Visual communication	Purpose(s)	Explain how the visual communication achieves it purpose or multiple purposes
iStock.com/Alex_Bond Music graphic for online music stores and CD artwork		
Lumiere Art & Co Blue Cage Light by Lumiere Art & Co		

ISBN 9780170401791

Visual communication	Purpose(s)	Explain how the visual communication achieves it purpose or multiple purposes
Hali Rugs Magazine page for Hali rugs		
Ryan Wheatley Illuminated transport stop poster		

ISBN 9780170401791

Visual communication	Purpose(s)	Explain how the visual communication achieves it purpose or multiple purposes
Harvest Textiles Flyer for Harvest Textiles workshops		
Law Architects Computer rendering of school building		

ISBN 9780170401791

2 In the grid provided, research and identify a *specific* example of a design that addresses each purpose. Include a brief description, image or sketch. Don't forget that if you use images that are not your own, you must attribute your sources.

Design field	Purpose	Example
Industrial design	To facilitate	
	To transport	
	To display	
	To protect	
	To interact	
	To accommodate	

ISBN 9780170401791

Design field	Purpose	Example
Environmental design	To shelter	
	To engage	
	To accommodate	
	To make sustainable	
	To beautify	
	To decorate	

ISBN 9780170401791

Design field	Purpose	Example
Communication design	To advertise	
	To inform	
	To promote	
	To explain	
	To identify	

ISBN 9780170401791

Design field	Purpose	Example
Communication design (cont.)	To illustrate	
	To highlight	
	To decorate	
	To guide	
	To teach	

ISBN 9780170401791

Design field	Purpose	Example
Communication design (cont.)	To symbolise	
	To communicate	
	To attract	
	To persuade	

ISBN 9780170401791

Chapter 4

CONTEXTS OF VISUAL COMMUNICATIONS

Where (location) and when (time) a design is used has a significant impact on its appearance, materials, format and content. Chapters 5 and 12 of *Nelson Visual Communication Design VCE Units 1–4* will help you with this chapter.

TASKS

1   Using the two very different contexts pictured – a business card and a billboard – create an identity for Beautiful Butterflies Pty Ltd. Your designs should have a unified visual relationship with one another, but should be appropriately adapted for each context.

(Copy or photocopy the butterflies and type to create your designs.)

Business card

Billboard

ISBN 9780170401791

BEAUTIFUL BUTTERFLIES PTY LTD

BEAUTIFUL BUTTERFLIES PTY LTD

ISBN 9780170401791

2 Research the designs pictured here and establish the designer (if known) and the date created. Investigate the social, cultural and technological contexts of each design. Describe how each design reflects the time in which it was designed. You can find colour versions of these images in Appendix III (page 205).

Shutterstock.com/Leonard Zhukovsky

Courtesy State Library of Victoria, with permission of The Continental

3 Pictured is the logo for the 'Coffee Café'. Apply the logo to the range of 'carriers' pictured on the following pages, taking the different contexts of each item into consideration. Create an eye-catching and engaging series of designs that remains consistent but conform to the appropriate context. You can find a colour version of the logo in Appendix III (page 205).

coffee café

ISBN 9780170401791

Company app for ordering coffee to pick up

Apron to be worn by barista and other staff

ISBN 9780170401791

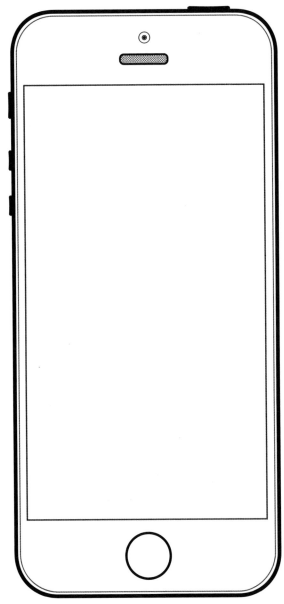

Mobile-friendly website with company info and
location details

Takeaway coffee cup

ISBN 9780170401791

Chapter 5
AUDIENCES

Understanding the target user/audience of a design is essential for a designer. The audience determines the success or failure of a design. User-centred and user-experience design are, increasingly, the core considerations of professional designers. Designing for the end user takes considerable research and thought to ensure that a design meets the needs and accounts for the abilities of its audience. Good design is about asking questions in the early stages of the design process to build a solid understanding of who the end user is. Understanding the characteristics of the audience is essential to designing products that are effective and meaningful.

You will find that Chapter 10 of *Nelson Visual Communication Design VCE Units 1–4* will help you with this section.

 TASKS

1 Target audience

Understanding the target user/audience of a design is essential for a designer.

Pictured on this page and next, are two 'audience collages' that visually describe two very different target audiences. Identify the characteristics (as relevant) of each audience based on the imagery used. You can find colour versions of these images in Appendix III (page 206).

iStockphoto.com: YanaZ, anderm, LiliGraphie, Leegudim, GoodLifeStudio. Shutterstock.com: Elena Sistaliuk, gomolach, LanaSweet, Wehands, Alessandro Colle, Everything.

AGE: _____

GENDER: _____

LOCATION: _____

SOCIOECONOMIC STATUS: _____

INTERESTS: _____

CULTURAL BACKGROUND: _____

ISBN 9780170401791

Prue Edmunds

AGE: _____

GENDER: _____

LOCATION: _____

SOCIOECONOMIC STATUS: _____

INTERESTS: _____

CULTURAL BACKGROUND: _____

2   Audience collage

Using the table on pages 218–19 of *Nelson Visual Communication Design VCE Units 1–4*, create an audience collage for one or more of the following design briefs. Identify clues in the description to assist in locating appropriate imagery. Remember to use correct copyright attributions for sourced materials (see pages 296–7 of *Nelson Visual Communication Design VCE Units 1–4*).

a   An electronic music festival aimed at the 18–30-year-old market. The festival is to take place at an island location. Participants will camp overnight. There will be art events, workshops, wellness seminars, healthy food options and chill-out zones. Tickets cost approx. $250 per person.

b   An easy-to-construct treehouse for use in suburban and rural backyards. The treehouse will be suitable for children aged 5–8 years of age. It will be built from sustainable materials and require minimal construction. A range of colourful and differently textured options are available to create a stimulating experience for the young user. The cost of the treehouse kit is $145.

c   The design of an interactive and engaging public space to house permanent museum exhibits. Focused on the natural world, the exhibits will feature native Australian flora and fauna. Families, students and school groups will be the most likely users to the space. Ideally the environment will be inviting and stimulating, offering visual, auditory and tactile activities. Entry is $8 for adults, children free.

3   Potential audiences

Pictured opposite are six potential audiences. On page 56 identify the features, interests, background, etc. of your chosen audiences and write a descriptive paragraph that summarises their key demographic characteristics. You can find colour versions of these images on pages 206–7.

ISBN 9780170401791

A   iStock.com/Eva-Katalin

B   iStock.com/michaeljung

C   Shutterstock.com/Monkey Business Images

D   iStock.com/Ximagination

E   iStock.com/max-kegfire

F   iStock.com/Yuri_Arcurs

ISBN 9780170401791

Selected audience   A   B   C   D   E   F   (Circle your selection)

a   What descriptive terms could be used to describe the age or age range?

_____

_____

_____

b   List adjectives that describe this audience.

_____

_____

_____

_____

c   What might be important to this audience?

_____

_____

_____

d   List interests, hobbies, lifestyle and activities that might be undertaken by this target audience.

_____

_____

_____

e   List words to describe possible socioeconomic factors that might affect this audience.

_____

_____

_____

_____

f   In your own words, write a summative description of the audience.

_____

_____

_____

_____

_____

_____

_____

_____

_____

_____

4 Persona design

Using the information on page 221 of _Nelson Visual Communication Design VCE Units 1–4_, create a persona for a typical user of each of the following products. The persona should include information about the interests, age, location, personality and background of a 'typical' user. You can find colour versions of these images on page 208.

ISBN 9780170401791

Persona A:

iStock.com/freestylephoto

Persona B:

Persona C:

5 Audience needs

 Pictured is an illustration of a standard screwdriver. Copy the image provided and create three alternative designs for the screwdriver handle for the audiences listed. Consider the needs of each of the audiences and ergonomic factors such as comfort, grip and accessibility.

a Design a handle for an elderly user who has limited flexibility due to arthritis. The user may need additional cushioning and/or the ability to rotate the handle for added comfort.

b Design a handle for a young child's first toolkit. The handle should suit the scale and proportion of a child's hand. The design may feature additional visuals and design elements such as colour and texture to add further appeal.

c Design a handle for an audience who are concerned with issues such as sustainability and waste reduction. They may be concerned about landfill and wish to buy quality tools that are durable and lasting for a lessened environmental impact.

Example

Shutterstock.com/Bonezboyz

ISBN 9780170401791

6 At the movies

Select a contemporary movie poster.

a Watch the trailer for the selected movie online.

b If required, read a synopsis of the plot.

c Respond to the following:

i What visuals in the poster provide clues to the movie's content/narrative?

_____

_____

_____

_____

ii Who is the likely target audience? Focus on age, gender, interests and location. Explain your answer.

_____

_____

_____

_____

_____

_____

iii What response does the poster aim to elicit from the audience (e.g. fear, intrigue, curiosity, sympathy, etc.)?

_____

_____

_____

_____

iv What design principles have been applied, and how do they convey the meaning/themes of the film? Describe the use of the design elements. How do they help to communicate meanings and themes to the audience?

_____

_____

_____

_____

_____

_____

_____

_____

_____

_____

7 Audience characteristics

Populate the table on the following page with as many descriptive terms and categories you can think of. This will build a broad vocabulary for you to use when identifying and describing the audience. See example:

Example

Characteristic	Descriptive terms
AGE descriptors	Elderly, middle-aged, youthful, children, adults, families, retirees, tweens, young adults, teenagers…

ISBN 9780170401791

Characteristic	Descriptive terms
Age descriptors	
Gender descriptors	
Socioeconomic factors	

ISBN 9780170401791

Characteristic	Descriptive terms
Interests	
Cultural & religious background	
Location	

ISBN 9780170401791

PART B

Chapter 6

TWO-DIMENSIONAL DRAWING

Chapter 1 of *Nelson Visual Communication Design VCE Units 1–4* will help you with this chapter.

Two-dimensional drawings provide a clear means of communicating information about the appearance, assembly, function or construction of an object. The methods of two-dimensional drawing used in the design field of industrial design are packaging nets and orthogonal drawings. In environmental design, plans and elevations are used.

TWO DIMENSIONAL DRAWING

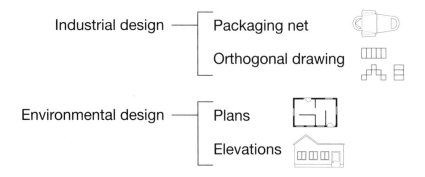

6.1 PACKAGING NETS

A packaging net is a two-dimensional drawing method used when an object is to be manufactured from a single piece of material; it is, as its title indicates, often used in packaging design. A packaging net provides information about the form of an object to be created; the application of different lines describes necessary folds and cuts.

 TASKS

1 Identify which three-dimensional object corresponds with the packaging net.

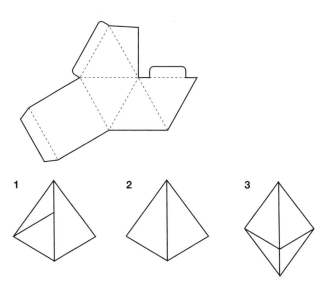

ISBN 9780170401791

2 Identify which packaging net corresponds with the three-dimensional object.

 1

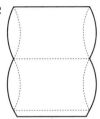 **2**

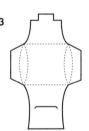 **3**

3 Match the correct three-dimensional cube to each drawing.

1

A B C

2

A B C

3
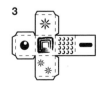
A B C

4 Match the development drawing with the correct image of it in an assembled state for the following two images. Circle the letter (A–D) beneath the correct image.

i

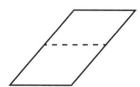

  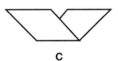
A B C D

ii

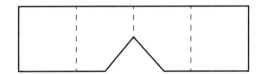

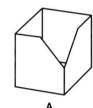   
A B C D

ISBN 9780170401791

TAKEAWAY

1 Collect examples of packaging at home or at your local supermarket. Unfold the packaging and redraw indicating all folds and tabs.

2 Design a die. Indicate the value of each face of the die without using numbers or dots. See example below.

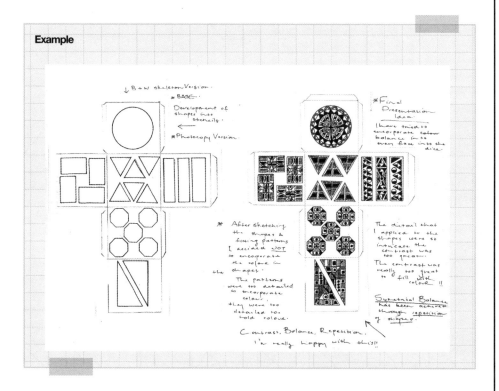

Example

Orthogonal drawing

Orthogonal drawings provide visual information about complex objects that cannot be constructed from a single piece of material. Designers, engineers, builders, architects and manufacturers use orthogonal drawings to specify the precise details of objects to be constructed or manufactured. Orthogonal drawings show several views of an object within the one drawing and conform to prescribed standards.

 TASKS

1 Select the correct views on the following page. Indicate by circling the letter that sits below the correct third-angle projection.

ISBN 9780170401791

1 Select the correct right-hand side view.

Front

A　　　　B　　　　C

2 Select the correct front view.

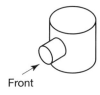
Front

A　　　　B　　　　C

3 Select the correct top view.

Front

A　　　　B　　　　C

4 Select the correct front view.

Front

A　　　　B　　　　C

2 Select the correct set of views. Indicate by circling the letter that sits above the correct third-angle projections.

1　　　　　　A　　　　　B　　　　　C

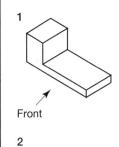
Front

2　　　　　　A　　　　　B　　　　　C

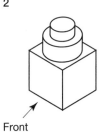
Front

3　　　　　　A　　　　　B　　　　　C

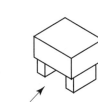
Front

4　　　　　　A　　　　　B　　　　　C

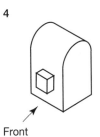
Front

ISBN 9780170401791

3   Draw the third-angle projection symbol in the space provided. Use the VCAA's *Technical Drawing Specifications*.

Third-angle projection symbol

Tom Grech

1   To familiarise yourself with drawing using only two dimensions, create a series of small orthogonal drawings of the following objects, showing three regular views. Use the minimum number of lines to convey information about the object. Remember to align the three views correctly. Add colour and tone if you would like to. You can find the colour version of the Coca-Cola image on page 208.

- washing machine
- church
- fish bowl
- egg on toast
- running shoe
- bowl of spaghetti
- hamburger
- key
- toaster
- Coca-Cola bottle
- double power point
- milk carton
- cupcake with a candle

Tom Grech

ISBN 9780170401791

Hidden detail

 TASKS

1   Using the correct line conventions, indicate where hidden details should be shown.

1

2

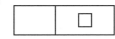

3

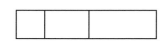

4

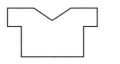

5

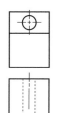

6

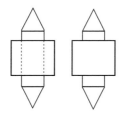

2   Draw the following three-dimensional objects as third-angle projection orthogonal drawings. Draw the FRONT, TOP and RIGHT-HAND SIDE views, including hidden detail. Draw your images at a scale of 2:1.

1

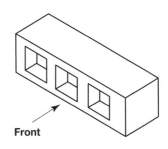

Front

2

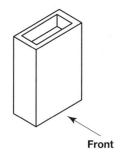

Front

3

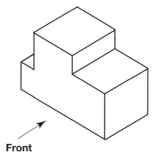

Front

4

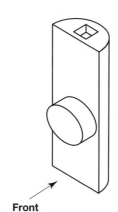

Front

5

Front

6

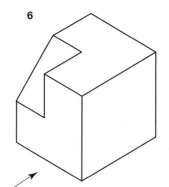

Front

ISBN 9780170401791

Dimensioning

Dimensioning is the placement of measurements on an orthogonal drawing. Like other aspects of this drawing method, there are strict conventions to be followed when dimensioning. You should refer to the VCAA's publication *Technical Drawing Specifications 2009*.

 TASKS

1 Add the missing centre lines to the orthogonal drawings. Apply appropriate line conventions to indicate the missing centre lines.

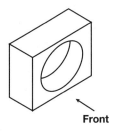

Front

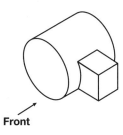

Front

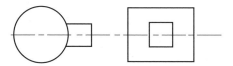

2 Complete the missing dimensions and labels on the following orthogonal drawings. Some dimensions have been provided for you. All dimensions are in millimetres.

1

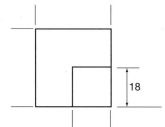

18

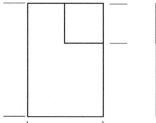

2 Ø 20

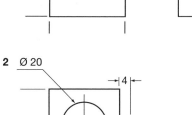

4

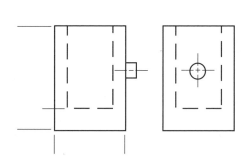

ISBN 9780170401791

3 Correctly dimension and label the following orthogonal drawings. Most dimensions have been provided for you. All dimensions are in millimetres.

1

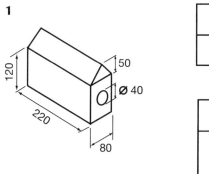

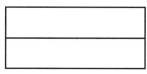

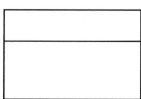

2

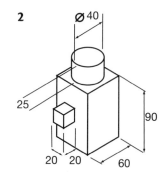

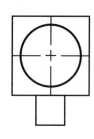

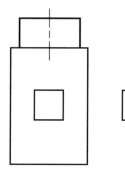

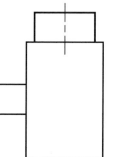

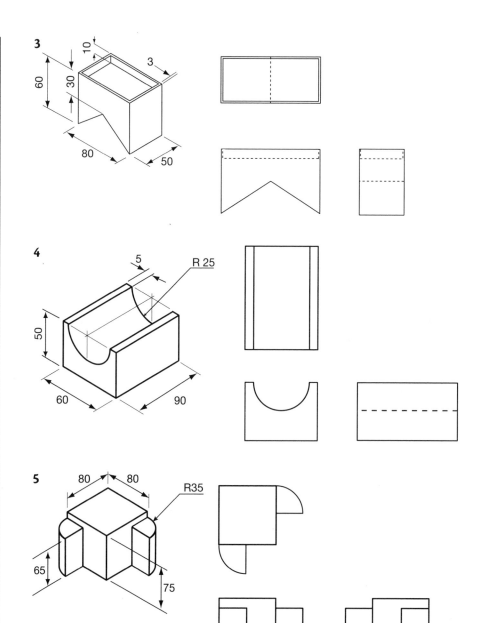

ISBN 9780170401791

4 Correctly dimension the circular details on the drawings below. Add centre lines, labels and hidden detail lines where required.

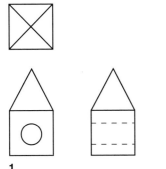

1
WIDTH: 25 mm
HEIGHT: 50 mm
DEPTH: 25 mm
DIAMETER OF CIRCLE: 10 mm

2
HEIGHT: 28 mm
DIAMETER OF CIRCLES: 30 mm, 24 mm, 17 mm

5 Draw these three-dimensional objects using the orthogonal drawing method. You should draw three regular views. Add dimensions, include labels, centre lines and hidden detail where required.

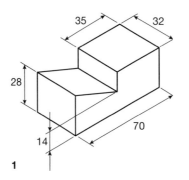

1

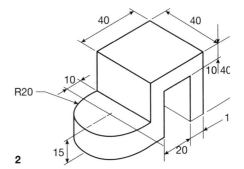

2

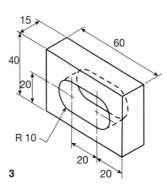

3

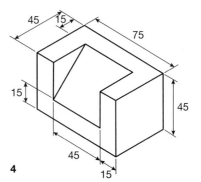

4

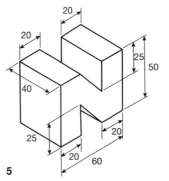

5

ISBN 9780170401791

Scale

Every orthogonal drawing must be drawn in proportion to the original three-dimensional object. The scale must be applied consistently throughout the drawing.

 TASKS

1 Draw the following isometric object as an orthogonal drawing at a scale of 1:10. Dimensions have been provided for you. All dimensions are in millimetres.

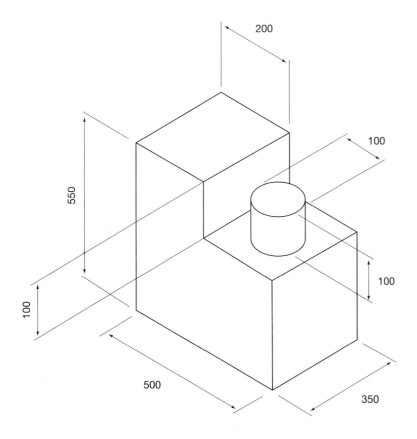

2 Draw the following isometric object as an orthogonal drawing at a scale of 2:1. Dimensions have been provided for you. All dimensions are in millimetres.

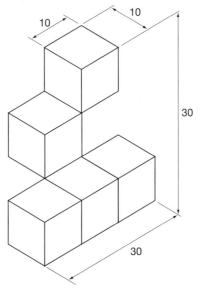

3 Draw the following isometric object as an orthogonal drawing at a scale of 1:1. Dimensions have been provided for you. All dimensions are in millimetres.

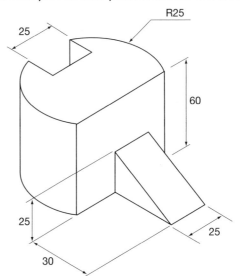

5

ISBN 9780170401791

Lettering

Instrumental drawings are created with the assistance of instruments such as rulers, set squares and other drawing instruments. They usually conform to set standards and follow rules and conventions – even about how labels look – to ensure consistency. When creating instrumental drawings, it is recommended that you apply a consistent text style to your labelling and annotation. VCAA's *Technical Drawing Specifications* requires that all views on an orthogonal drawing be labelled appropriately with VIEW labels; for example, FRONT VIEW, TOP VIEW, SIDE VIEW.

When labelling your views (or adding any other annotations to your instrumental drawings) you should use upper-case (capital) letters in a sans serif typeface. (See Chapters 5 and 6 of *Nelson Visual Communication Design VCE Units 1–4* for more information about typefaces.)

 TASKS

1 Use the guidelines below to practise your lettering. (When you are drawing manually, you may choose to use guidelines to assist you in labelling appropriately.) Aim to write labels that are around 5 millimetres in height (approximately 14 points).

FRONT VIEW   **TOP VIEW**   **SIDE VIEW**

FRONT VIEW   TOP VIEW   SIDE VIEW

ISBN 9780170401791

Cross-sections

 TASKS

1   Complete the missing details on the cross-section views.

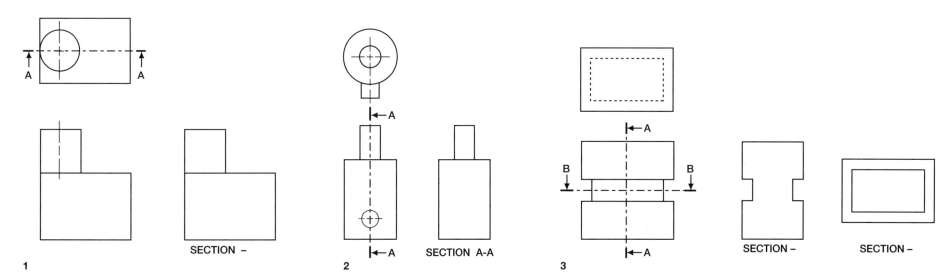

SECTION –

SECTION A-A

SECTION –

SECTION –

1

2

3

ISBN 9780170401791

2 Match the correct cross-section with the appropriate orthogonal drawing. Label each of the section views appropriately; for example, SECTION A–A.

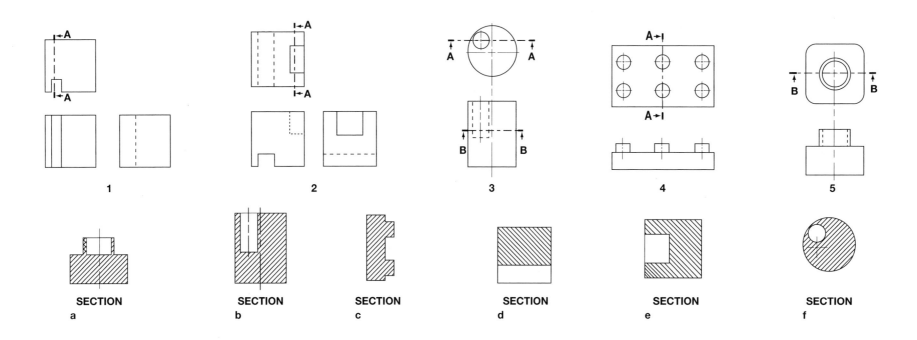

3 Draw cross-sections for these objects. The cutting plane has been indicated.

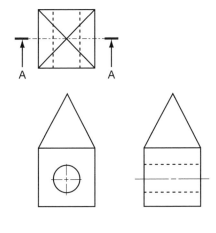

1

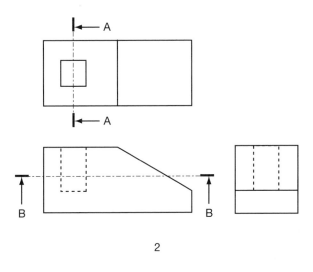

2

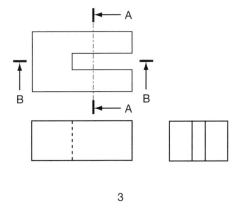

3

SECTION A-A

SECTION A-A   SECTION B-B

SECTION A-A   SECTION B-B

ISBN 9780170401791

Plans and elevations

In environmental design, plans and elevations are used to communicate visual information about a three-dimensional space or design. Plans are equivalent to the top view of an orthogonal drawing while elevations show the front and side views.

 TASKS

1   Create a plan view of a building you are familiar with such as your school, home or bedroom. Indicate the location of doors, windows, structural walls, etc. Use symbols to illustrate the purpose of each room.

   2   Create external elevations from the provided plan view. You may wish to photocopy and enlarge the plan then cut it out to use as the source of information to create your elevations.

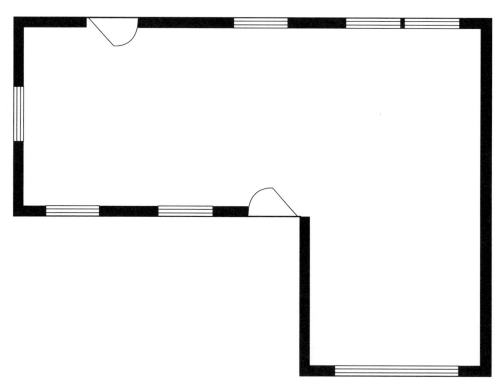

ISBN 9780170401791

3   A local early learning centre has commissioned you, a landscape architect, to design a new playground area for their young clientele.
The design brief requires the following:

- privacy from the main road

- north-facing play area with slide, swings, etc.

- areas for shade

- sandpit

- quiet space

- playground visibility from the early learning centre.
On the existing plan (pictured), create an appealing and
safe playground that also includes the above criteria.

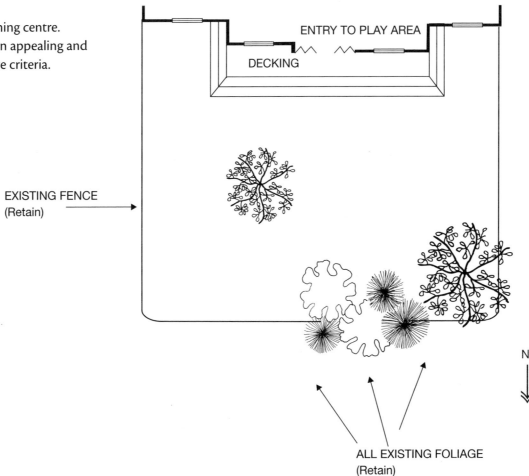

ENTRY TO PLAY AREA

DECKING

EXISTING FENCE
(Retain)

MAIN ROAD

N

ALL EXISTING FOLIAGE
(Retain)

ISBN 9780170401791

4 Label the following architectural symbols.

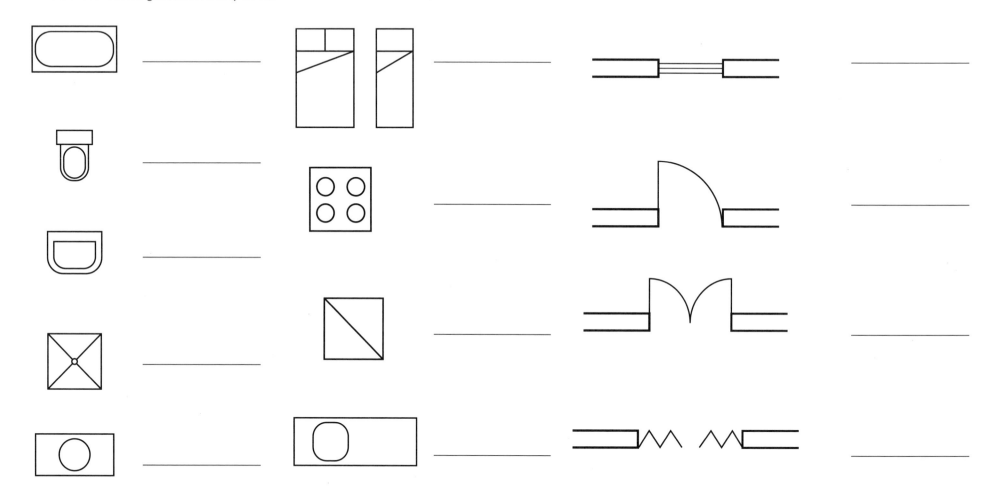

ISBN 9780170401791

5   Add dimensions to the pictured plan of a backyard studio/home office. Carefully consider and identify the most appropriate scale.

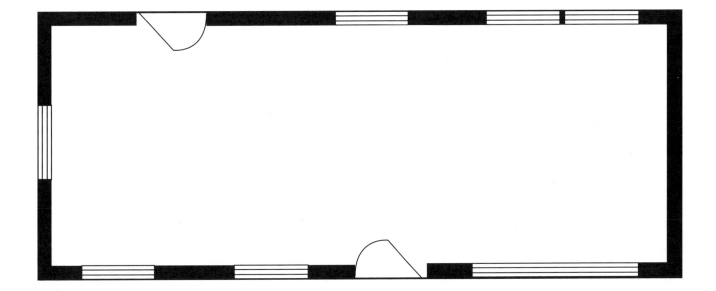

ISBN 9780170401791

Chapter 7

THREE-DIMENSIONAL DRAWING

Chapter 1 of *Nelson Visual Communication Design VCE Units 1–4* will help you with this chapter.

Three-dimensional drawing more clearly represents how we see objects, as we are accustomed to observing the length, width and depth of objects. In VCE Visual Communication Design there are several three-dimensional drawing methods covered in detail:

THREE DIMENSIONAL DRAWING

Paraline drawing methods —— Isometric

Planometric

Perspective drawing methods —— One-point perspective

Two-point perspective

7.1 PARALINE DRAWING

Isometric drawing

In isometric drawing the height (or corner) of the object faces the viewer and the width and depth of the object are drawn parallel at 30°.

TASKS

1. Draw the following simple objects using the isometric drawing method. Photocopy and enlarge. Measure to determine the dimensions.

1
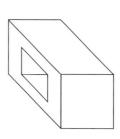

2
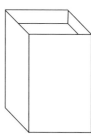

3
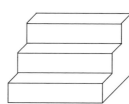

4

5
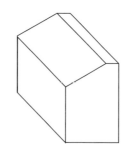

6
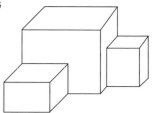

ISBN 9780170401791

2 Draw the following orthogonal drawings in the isometric method. Photocopy and enlarge to 200%. Measure to determine the dimensions.

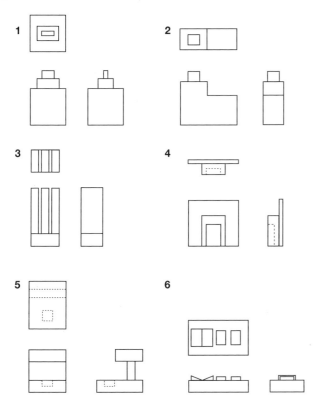

3 Redraw the exploded isometric view as a whole isometric drawing.

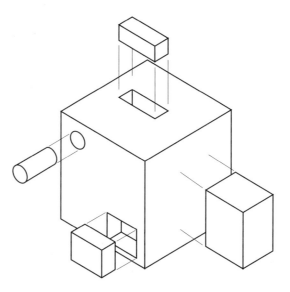

4 Use isometric grid paper (Appendix I, page 189) to create an exploded isometric view of the recycling bin (the lid and wheels are removable).

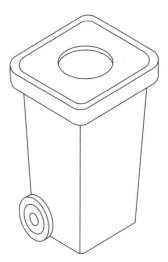

ISBN 9780170401791

5 Draw these orthogonal views of products as isometric drawings.

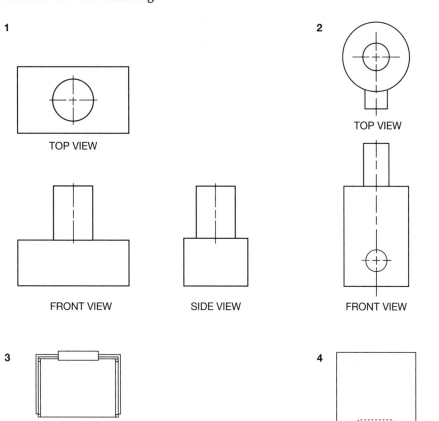

1

TOP VIEW

FRONT VIEW   SIDE VIEW

2

TOP VIEW

FRONT VIEW

3

TOP VIEW

FRONT VIEW   SIDE VIEW

4

TOP VIEW

FRONT VIEW   SIDE VIEW

ISBN 9780170401791

Planometric drawing

In planometric drawing the height (or corner) of the object faces the viewer and the width and depth of the object are drawn parallel at 45°. Planometric drawings are commonly used in interior design and architecture.

 TASKS

1 Draw the following orthogonal drawings as planometric drawings.

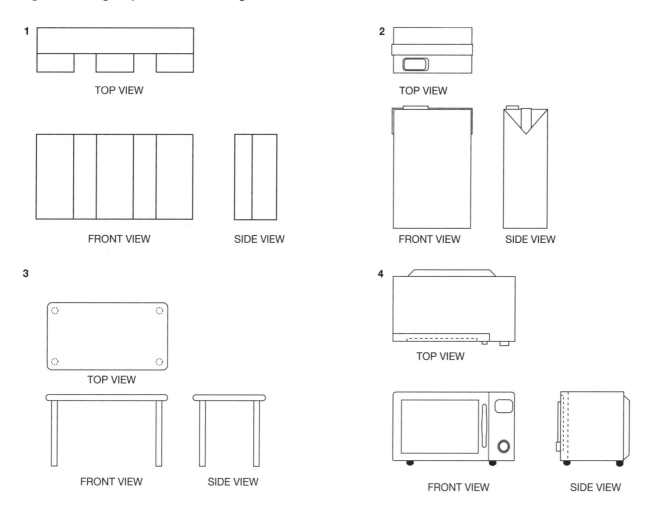

ISBN 9780170401791

2 Draw the plan views in the planometric drawing method. Use the example as a guide. Measure to determine dimensions (round to the nearest millimetre).

1

Table and chairs

Counter

Window

Draw from this direction

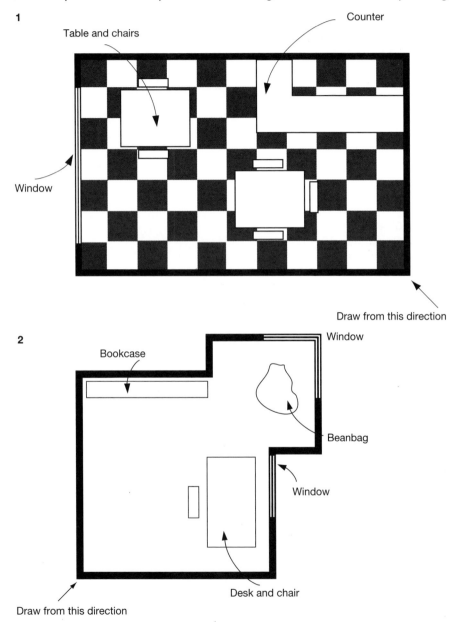

2

Bookcase

Window

Beanbag

Window

Desk and chair

Draw from this direction

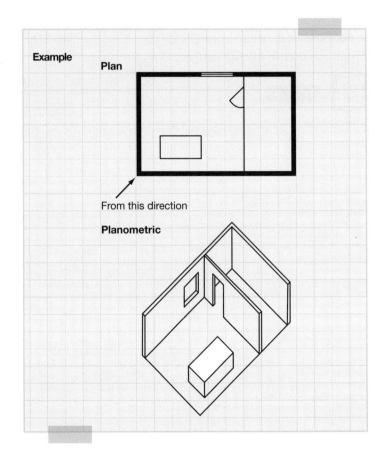

Example

Plan

From this direction

Planometric

ISBN 9780170401791

3 Draw the isometric objects (below) using the planometric drawing method. Measure to determine dimensions (round to the nearest millimetre).

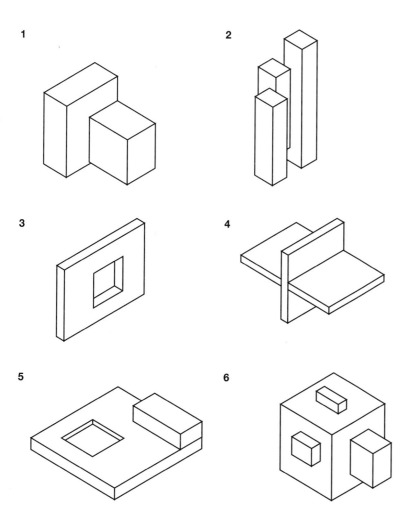

Ellipses

 **TASKS**

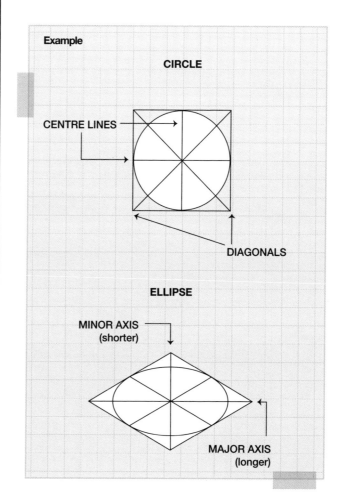

1   The 'falling pancakes' exercise below will help you to practise freehand ellipses. Use the guidelines to create freehand ellipses of different sizes. Increase the diameter of the ellipse as it 'falls' to the bottom of the grid (see example).

Example   20 mm ellipses   40 mm ellipses   50 mm ellipses   75 mm ellipses

2 Draw ellipses onto the following paraline forms (see the example on page 85). Use the major and minor axes to assist your drawing. Remember that the longest part of the ellipse follows the major axis. Add axes where required. On objects 3 and 7, add an additional ellipse to form a hollow cylinder.

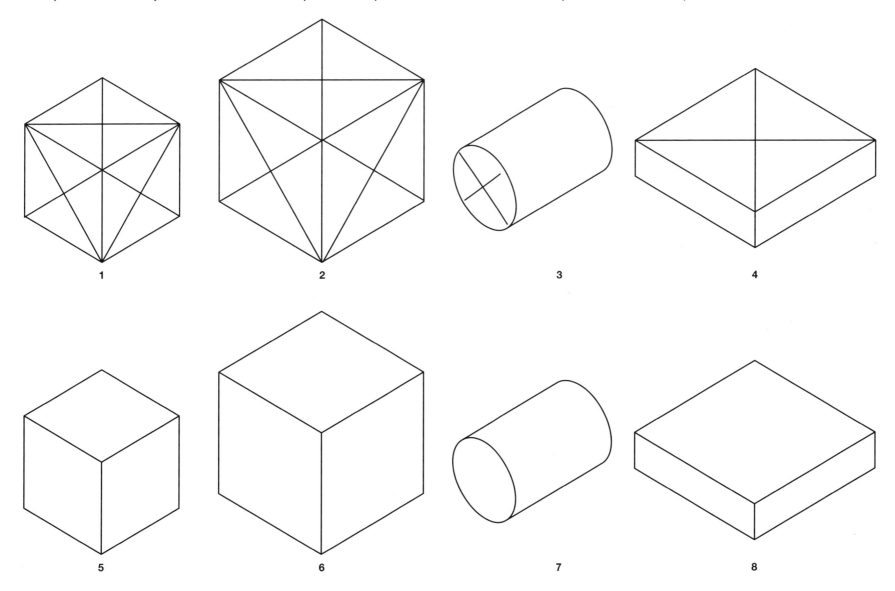

1

2

3

4

5

6

7

8

ISBN 9780170401791

3 Draw ellipses onto the perspective forms. Use the major and minor axes to assist your drawing. Remember that the longest part of the ellipse follows the major axis.

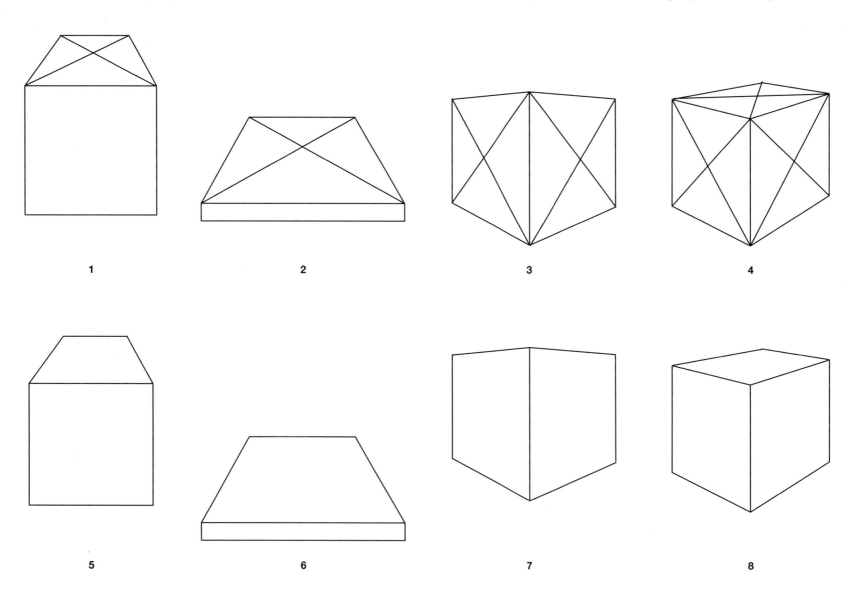

1

2

3

4

5

6

7

8

ISBN 9780170401791

7.2 PERSPECTIVE DRAWING

In any perspective drawing, how an object is placed in relation to the horizon line affects the point of view of the depicted object. The horizon line sits at the level of the viewer's eyes. This is called the eye level. An object placed below the horizon line – below eye level – gives the viewer more information about the top of the object. An object placed above the horizon line provides visual information about the underneath area. An object placed directly on the horizon line will create a realistic eye-level view.

One-point perspective drawing

One-point perspective is sometimes referred to as linear perspective. In one-point perspective an entire plane of an object faces the viewer.

Key concepts

Key concepts to remember when drawing in one-point perspective:

- the height and width of the object face the viewer

- all depth (or the sides of the object) recedes to one point on the horizon line.

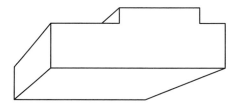

 TASKS

1 Observe the following one-point perspective drawing. Using a ruler or freehand, identify where the vanishing point and horizon line are situated and draw them.

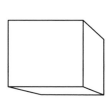

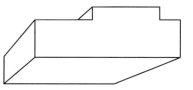

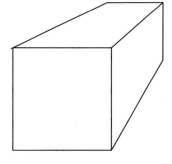

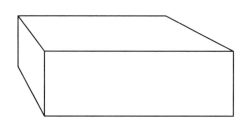

2 Add five simple objects of your own using the same vanishing point. (The depth of the object towards the horizon line is not specified.)

ISBN 9780170401791

3 Complete the objects into their three-dimensional perspective forms using the provided horizon line and vanishing point. (The depth of the object towards the horizon line is not specified.)

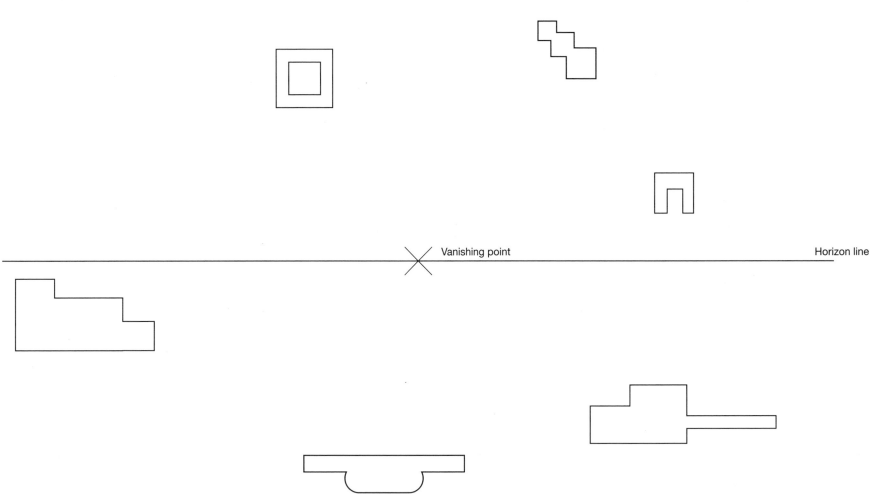

Vanishing point

Horizon line

ISBN 9780170401791

4 The current illustration is drawn at eye level. Redraw the illustrated object twice, once in a position above eye level and once in a position below eye level. The depth of the object towards the horizon line should be in proportion to the illustration provided.

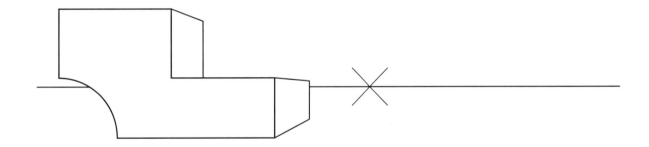

ISBN 9780170401791

5   Using the example as inspiration, add detail to the room interior on page 93 using the provided vanishing point and horizon line. You may use the provided images as inspiration or create your own.

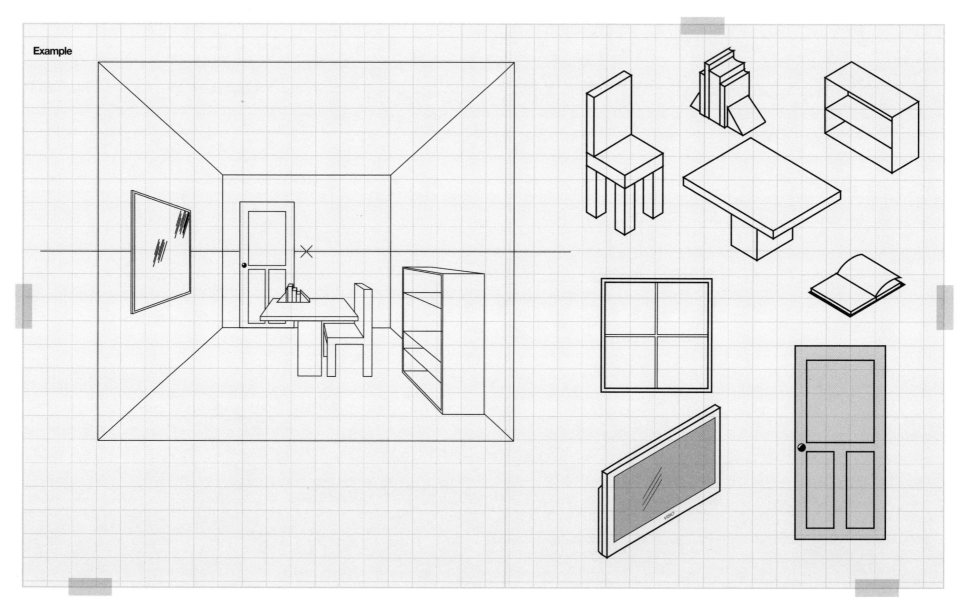

Example

ISBN 9780170401791

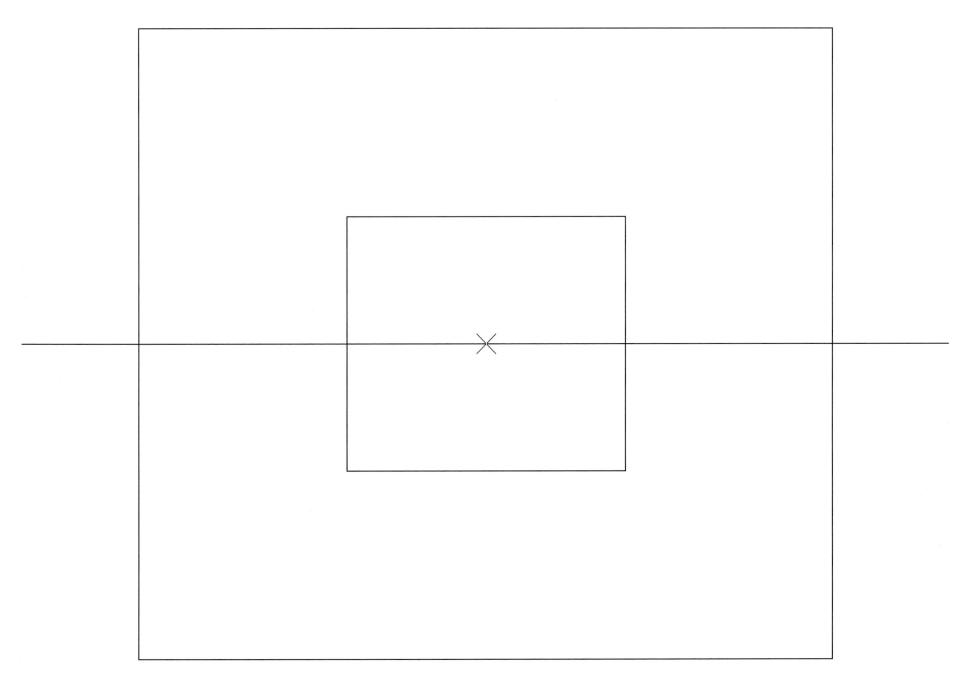

ISBN 9780170401791

TAKEAWAY

1   Create a one-point perspective view of your bedroom from an unusual angle; for example, the ceiling looking down or looking through a window. Include essential details such as doors, furniture and windows.

Two-point perspective drawing

Two-point perspective is sometimes referred to as angular perspective. In two-point perspective only the height faces the viewer and the depth or sides of the object recede to two vanishing points on the horizon line.

Key concepts

Key concepts to remember when drawing in two-point perspective are:

*   the height of the object faces the viewer

*   all other dimensions recede to two points on the horizon line.

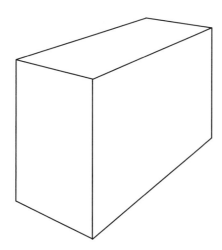

 TASKS

1   Look at the following two-point perspective drawing. Using a ruler or freehand, identify where the vanishing points and horizon line are situated and draw them.

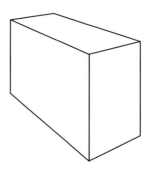

2   Add five simple objects of your own using the same vanishing points. (The depth of the object towards the horizon line is not specified.)

ISBN 9780170401791

3   Redraw these solid forms in two-point perspective (as shown) then transform them to open boxes. Add flaps or other details for greater complexity (see example). You may work at your own scale or one indicated by your teacher.

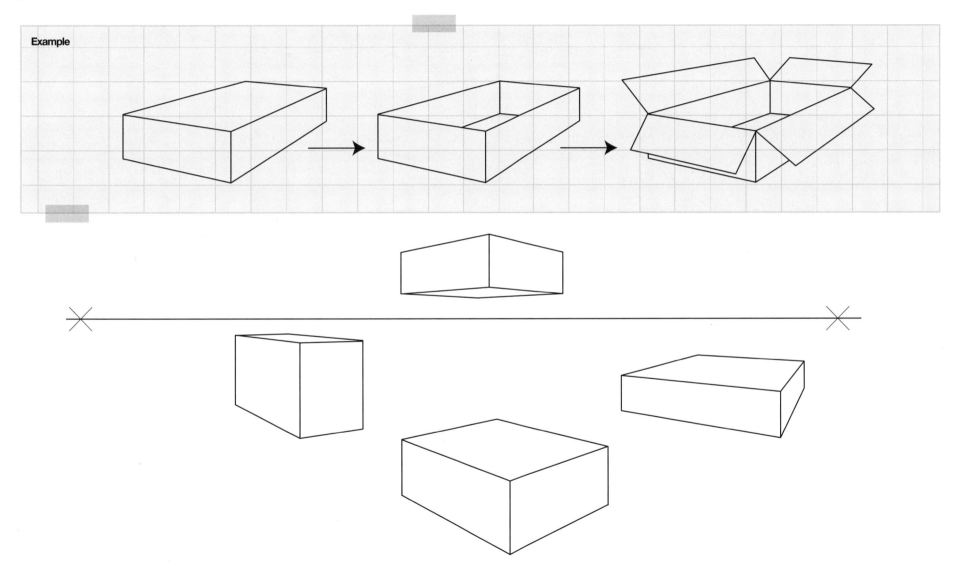

ISBN 9780170401791

4 Find and indicate the perspective centres of each plane (side) of the following objects (see example). Use diagonal lines to establish the perspective centre of a plane.

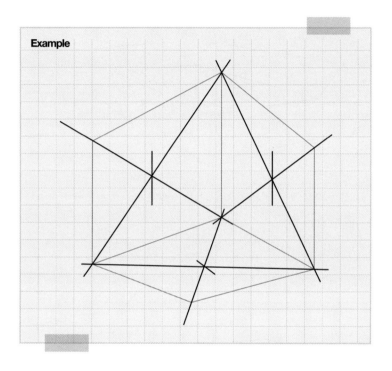

Example

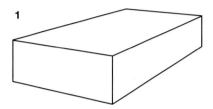

1

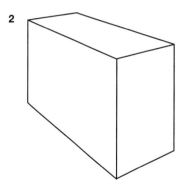

2

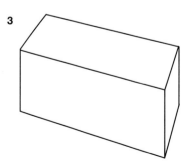

3

ISBN 9780170401791

5 Find the centre points of the following objects and add the provided detail (see example).

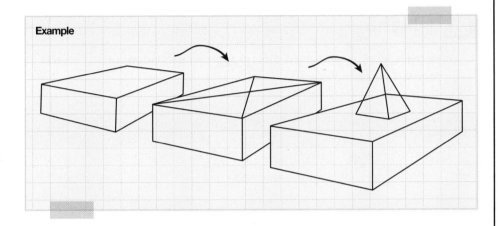

Example

1

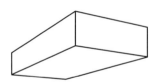

Pyramid detail in centre of bottom plane

2

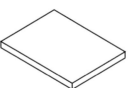

Rectangular detail in centre of front plane

3

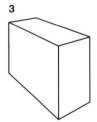

Square detail in centre of side plane

6 Using the plan view and elevation provided, create a two-point perspective drawing of the dwelling using the techniques outlined in *Nelson Visual Communication Design VCE Units 1–4*, page 15.

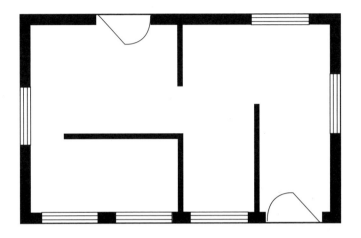

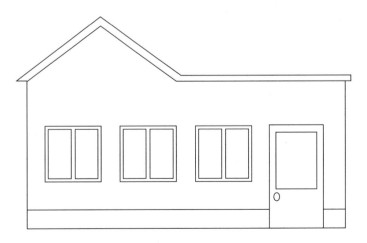

ISBN 9780170401791

TAKEAWAY

1   Create a two-point perspective building with the features:

- steps
- chimney
- doors
- windows
- veranda
- pitched roof with dormer window.

Or create a building of your own (see example).

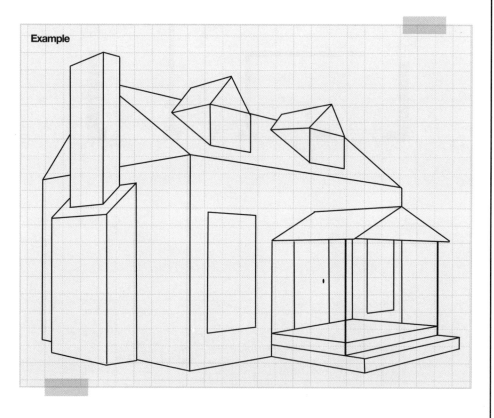

Example

2   Create a two-point perspective city-scape using only 0.4 mm and 0.8 mm fineliners. Create a dramatic light source and render in black and white to create a comic book or graphic novel appearance to your work.

Drawing shadows

TASKS

1   Using the light source indicated, draw in appropriate shadows on the objects on page 99 (see example).

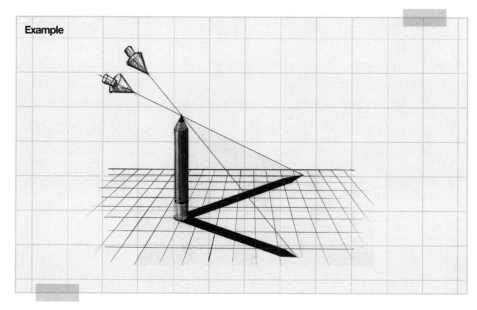

Example

ISBN 9780170401791

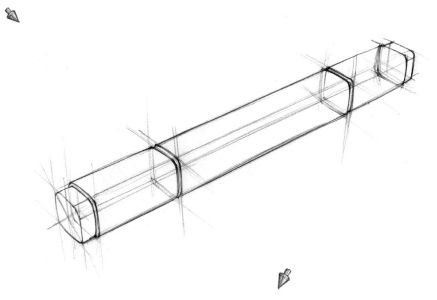

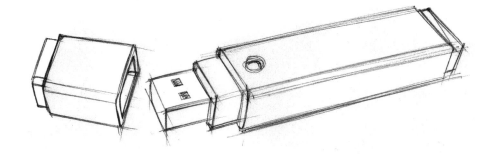

ISBN 9780170401791

Drawing complex objects

TASK

These perspective images have been left deliberately rough to enable you to see the helpful construction lines. Use these, and similar lines, in your drawing to ensure that you develop correct proportions and scale. Read pages 22–3 of *Nelson Visual Communication Design VCE Units 1–4* to assist you in using the crating technique in drawing construction. You can find coloured versions of the images in Appendix III (pages 209–10).

Using the construction images as a guide, recreate the original image of the objects on paper. It will help to identify the *horizon line* and *vanishing point* to ensure that your perspective drawing is accurate.

When you feel confident, render the image with suggested media and details:

- marker (if using a marker, use bleedproof paper)
- pencil
- dry pastel
- ink pen (e.g. a fineliner or technical pen)
- render to express surface texture
- shadow and highlights
- tonal variations
- render to indicate materials.

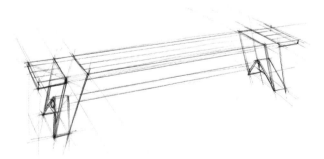
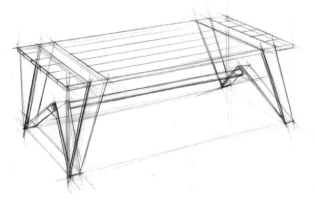

Mark Wilken (all on page)

Image 1: Outdoor setting

ISBN 9780170401791

Image 2: Coffee pot, mug and coffee jar

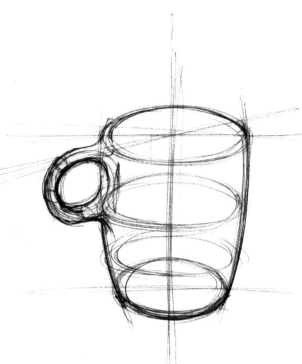

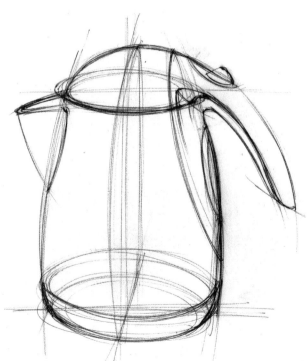

Mark Wilken (all on page)

ISBN 9780170401791

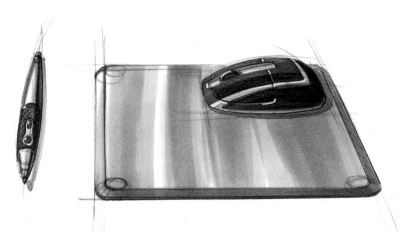

Image 3: Digital tablet, stylus & mouse

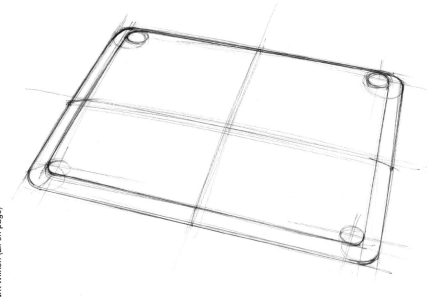

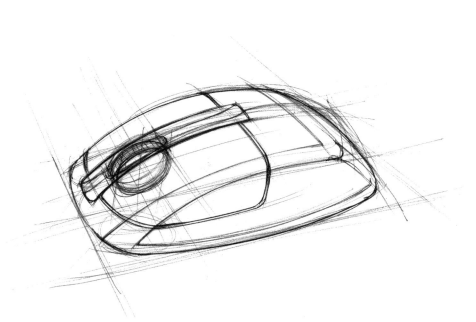

ISBN 9780170401791

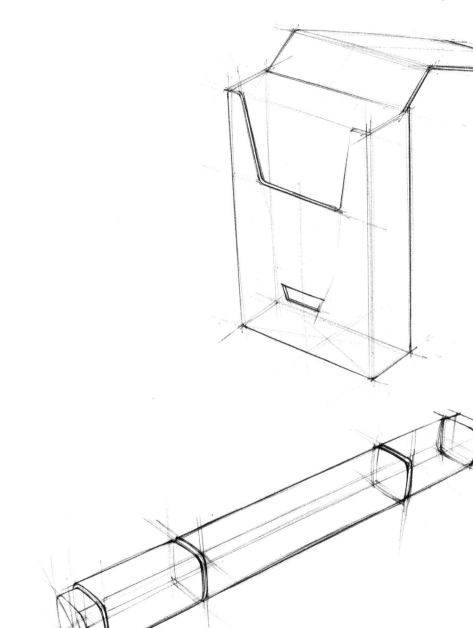

Image 4: Set of markers and packaging

ISBN 9780170401791

Mark Wilken (all on page)

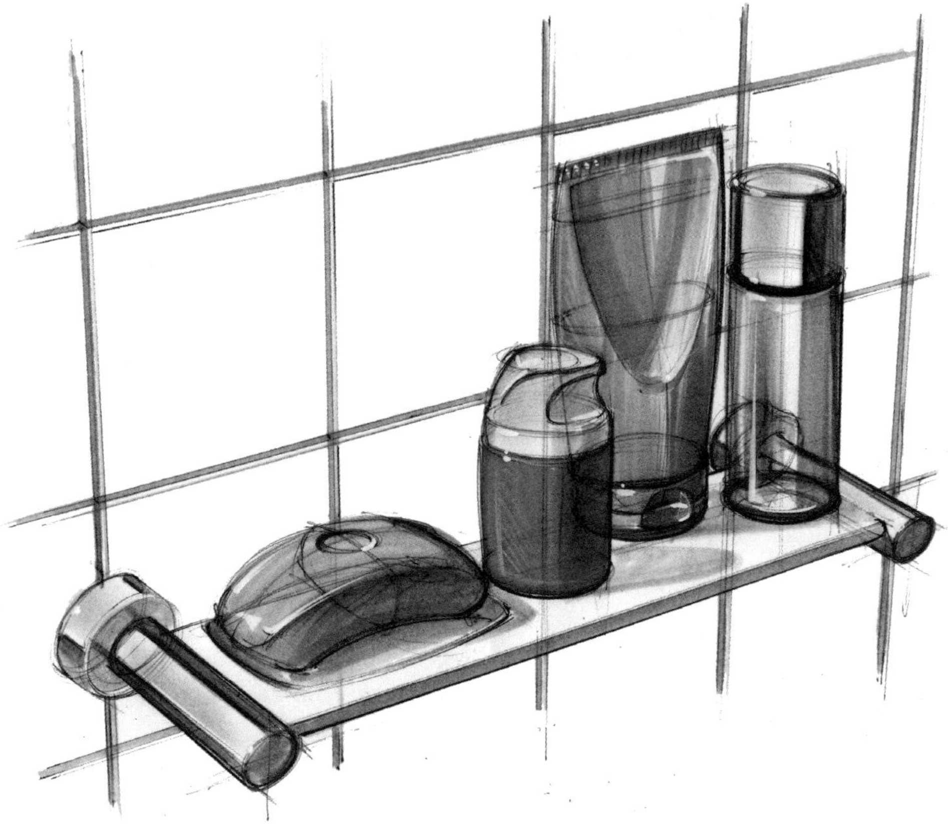

Image 5: Bathroom shelf with toiletry products

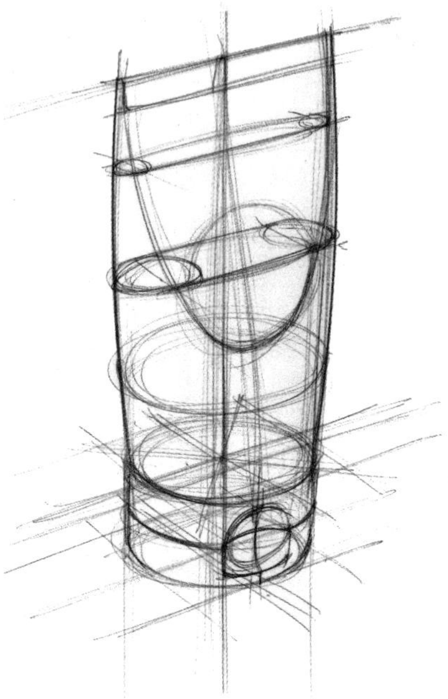

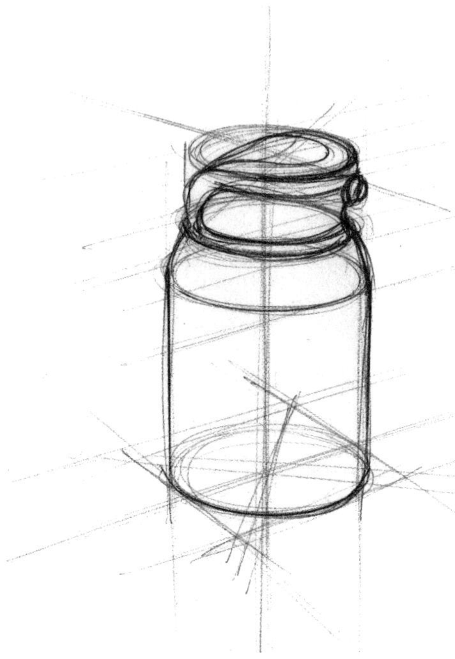

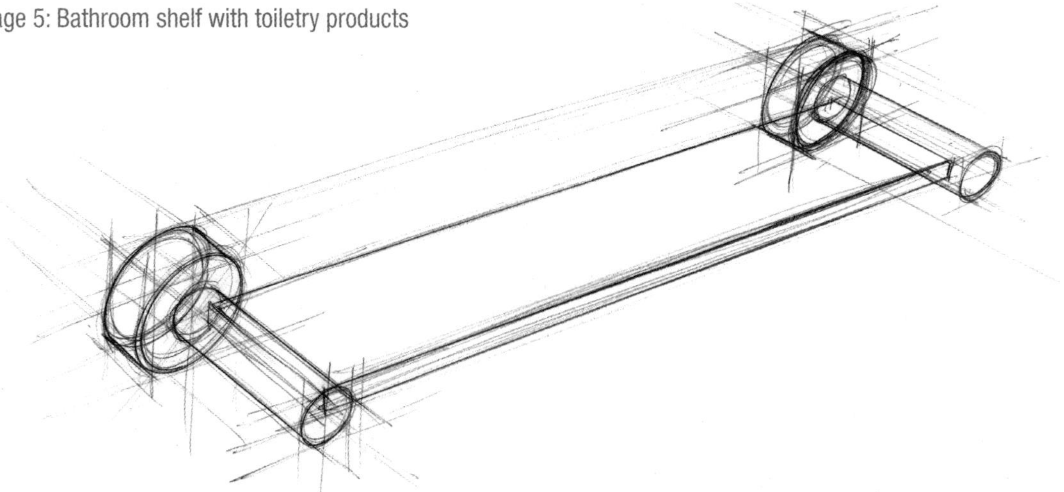

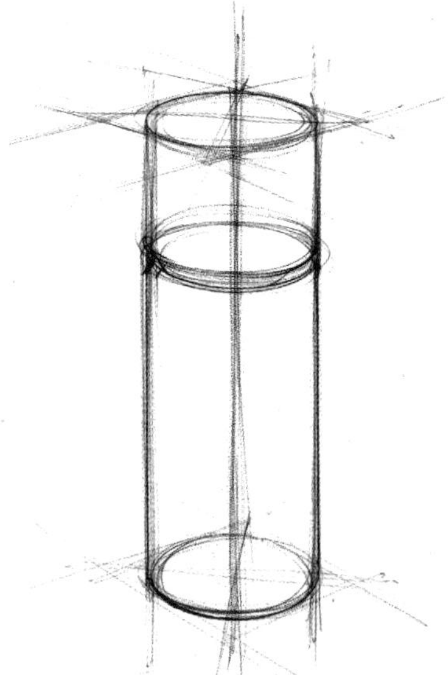

Mark Wilken (all on page)

Chapter 8
RENDERING TEXTURES & MATERIALS

This chapter is designed to help you build skills in rendering textures and materials using a range of media and techniques. You will use tone, line, colour and texture to emphasise the three-dimensional qualities of objects and to suggest textural details. Chapter 2 of *Nelson Visual Communication Design VCE Units 1–4* will help you with this chapter.

8.1 TONE

 **TASKS**

1 To practise your application of tone, complete the TONAL GRADIENT grid in a range of media. Start with the darkest tone and remember that your lightest tone should be paper.

DARK ⟶ MEDIUM ⟶ LIGHT

Greylead pencil 6B				
Coloured pencil				
Marker				
Fineliner (dot render)				
Fineliner (crosshatch)				

ISBN 9780170401791

2   On the following shapes, render using tones in reference to the given light source.

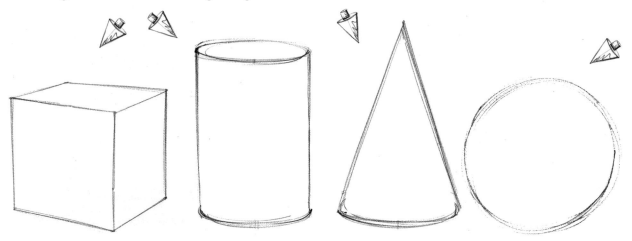

3   Now render using tones in reference to the given light source on these slightly more complicated shapes.

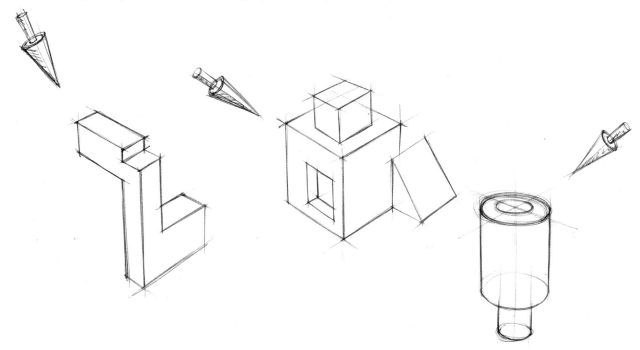

ISBN 9780170401791

8.2 TEXTURE

 **TASKS**

1   Using a sharpened, soft pencil (6B preferred) or another media of your choice, draw the following textures using the images as a guide to tones, shadows and highlights.

Images this page and next all author's own except iStock.com/Noi_Pattanan (water drop on leaf), iStock.com/zoran simin (metal rings) and iStock.com/ligora (tyres).

ISBN 9780170401791

2   Apply colour, texture and tone to render the following objects. Add detail to clearly show the indicated texture and materials.

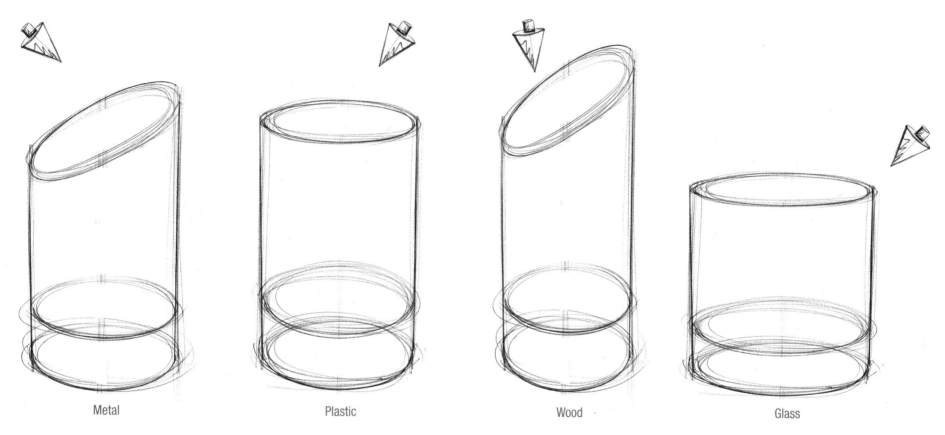

Metal   Plastic   Wood   Glass

 TAKEAWAY

1   Find a black-and-white photograph that uses strong contrasts between highlights and shadows (a portrait is a good choice). Cut away half of the image and stick it into your book; use a sharp 6B pencil to render it effectively. Use the cut-away portion of the image as a guide.

ISBN 9780170401791

8.3 RENDERING MATERIALS AND TEXTURES

Over the following pages you will find a range of images that are made from different materials. Using the rendered images as a guide, render the line drawings to represent form, texture and materials.

You may use any media or combination of media to complete the images. The rendered colour versions of these line drawings are in Appendix III (pages 210–15). Videos showing how the images were rendered will be on the *Nelson Visual Communication Design VCE Units 1–4* website. They were rendered using Copic markers, pastel and coloured pencil.

Representing natural materials

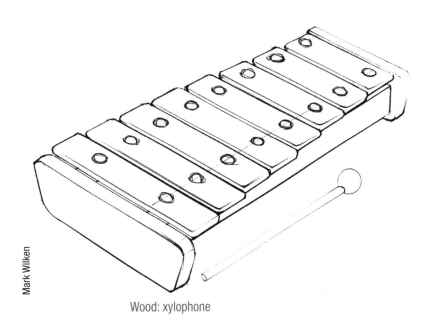

Mark Wilken

Wood: xylophone

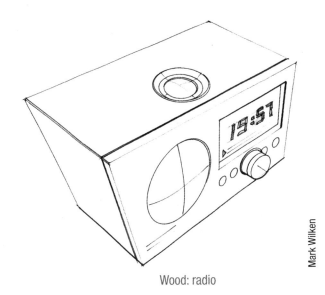

Mark Wilken

Wood: radio

Representing textiles

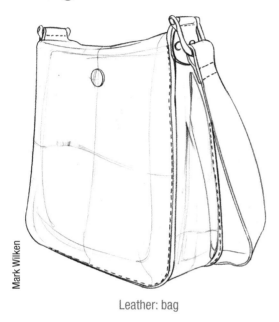

Mark Wilken

Leather: bag

Mark Wilken

Fabric: lampshade

Mark Wilken

Fabric: various

ISBN 9780170401791

Representing metallic and reflective materials

Mark Wilken

Glass: perfume bottle

Mark Wilken

Metal: Coffee pot

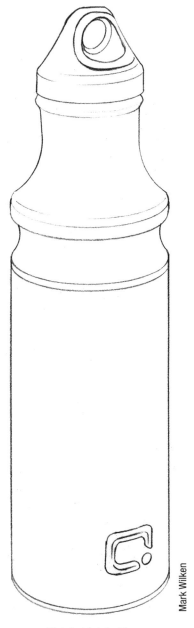

Mark Wilken

Metal: drink bottle

ISBN 9780170401791

Representing plastics

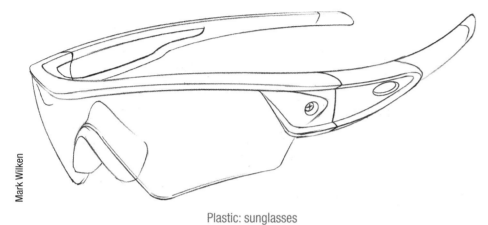

Mark Wilken

Plastic: sunglasses

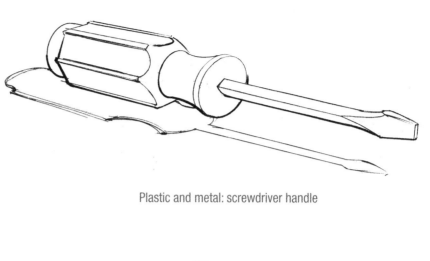

Mark Wilken

Plastic and metal: screwdriver handle

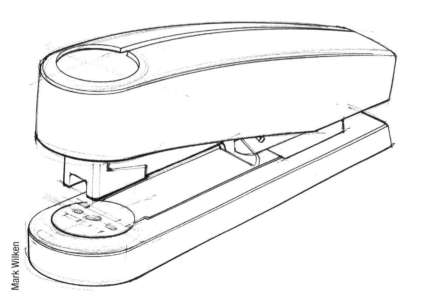

Mark Wilken

Plastic and metal: stapler

Mark Wilken

Plastic: coffee cup

ISBN 9780170401791

Representing eco materials

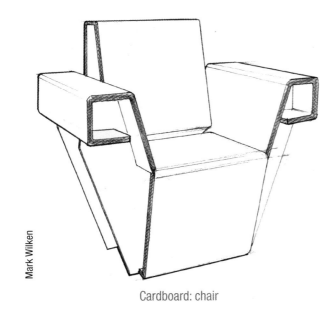

Mark Wilken

Cardboard: chair

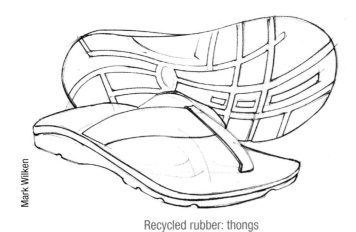

Mark Wilken

Recycled rubber: thongs

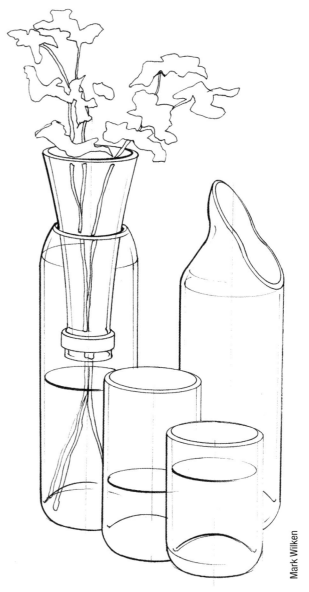

Mark Wilken

Recycled glass: 'Transglass' glassware

ISBN 9780170401791

Representing composite materials

Concrete: planter

Mark Wilken

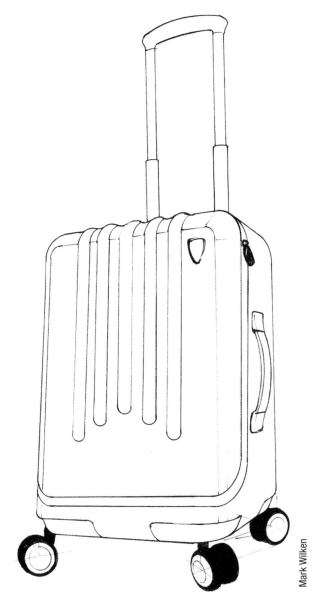

Carbon fibre: suitcase

Mark Wilken

ISBN 9780170401791

Representing ceramics

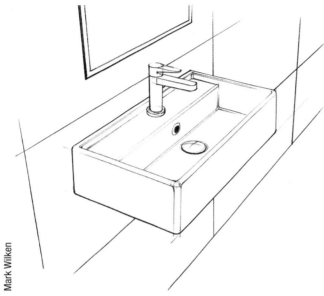

Ceramic: bathroom vanity

Ceramic: homewares

Representing combined materials

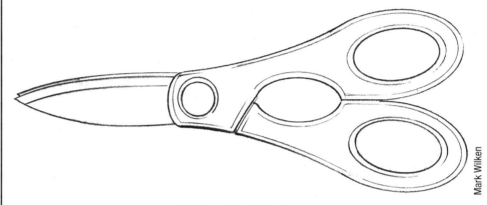

Rubber and metal: scissors

Fabric, leather and rubber: running shoes

ISBN 9780170401791

▣ EXTRA TASKS

1 Identify the materials you think have been used to make the office suite (the objects, not the drawing).

2 Render this image of the office desk using media of your choice. Pay attention to the characteristics of materials (e.g. transparent glass) and depict all relevant textural details.

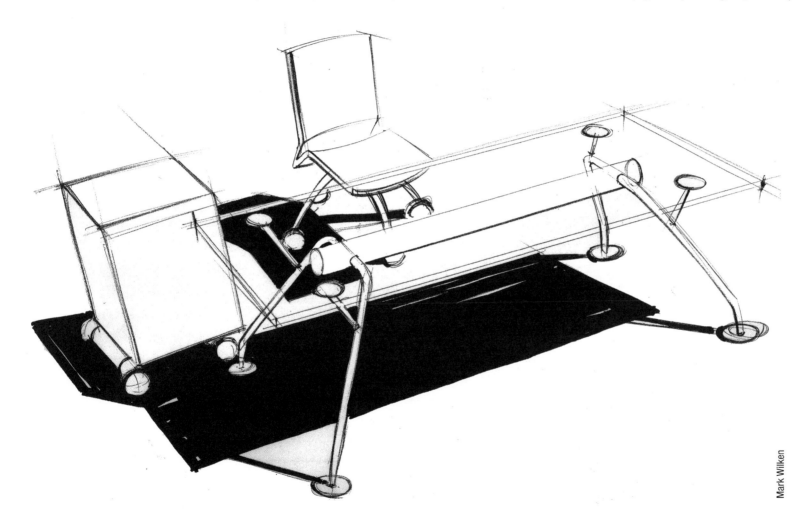

Mark Wilken

ISBN 9780170401791

3   Identify the materials you think have been used to make the contemporary kitchen (the objects, not the drawing).

4   Recreate the rendered contemporary kitchen using media of your choice. Pay attention to shadow areas and use the appropriate perspective method when applying linear details.

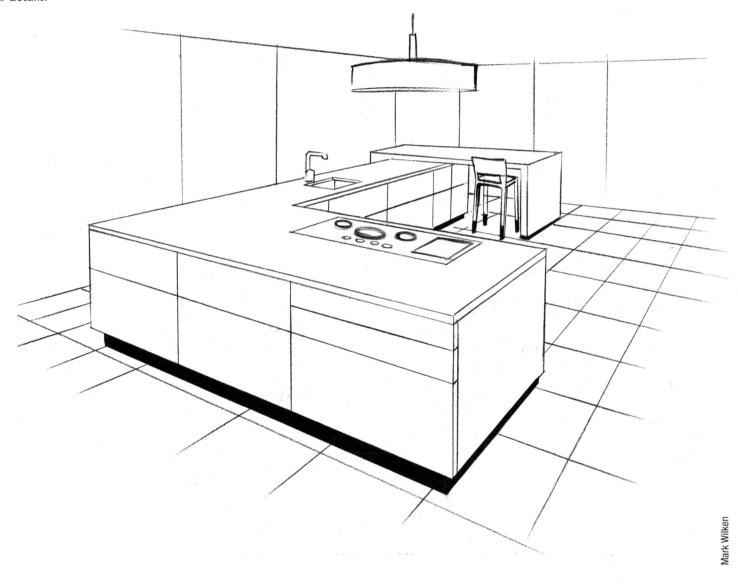

Mark Wilken

ISBN 9780170401791

TYPOGRAPHY & LAYOUT

Typography is an essential aspect of communication design. This chapter will assist you in understanding some of the terms and applications of effective typography. Chapter 6 of *Nelson Visual Communication Design VCE Units 1–4* will help you with this chapter.

9.1 THE LANGUAGE OF TYPOGRAPHY

It is important to familiarise yourself with the terminology used in typography. Identification and communication of typographic elements in analysis and design annotation are good skills to have in your Visual Communication Design toolbox.

TASKS

1 Identify and label the following typographic features.

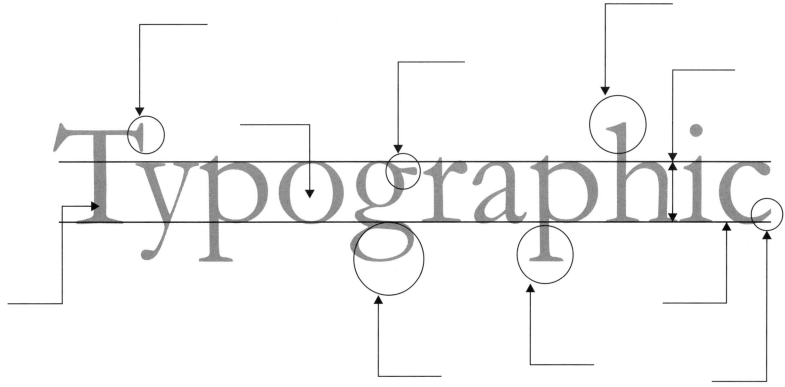

ISBN 9780170401791

2   Label each of the following serif styles.

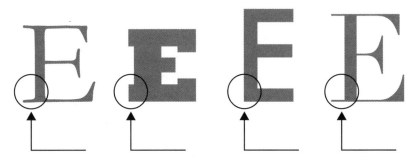

3   Match the appropriate term with the typographic feature.

Ligature	&
Sans serif	ff
Kerning	*F*
Ampersand	12345
Non-lining numerals	TYPE
Serif	**TYPE**
Decorative	Type
Swash	Type
Tracking	Type

4   Label the location of tracking/spacing, leading and kerning on the following example of type.

Flying
Fish

5   Create your own 'Typoscope' using a typeface of your choice. Combine glyphs (letterforms) to create a recognisable but abstract image.

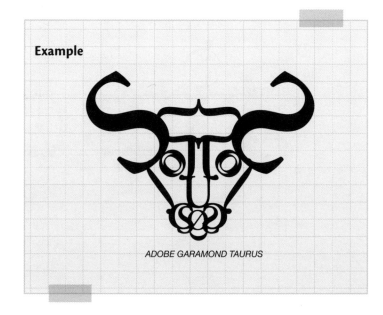

Example

ADOBE GARAMOND TAURUS

ISBN 9780170401791

9.2 TYPE IN PRACTICE

Whether using handwritten or digital type, it is important to understand the nature of letterforms and the relationships they hold with other type and with images. Practice, in this regard, really does make perfect.

 TASKS

1   The meaning of a word can be expressed through the spacing, placement and scale of type on a page. The use of type to convey meaning is seen often in editorial contexts, in advertising and on screen. The choice of typeface can convey a mood, idea or concept but so too can the placement of individual glyphs (letterforms).

Design a series of typographic images that represent a series of actions. Using *tracking/spacing*, *kerning*, *point size* and the arrangement of letterforms, create simple compositions that visually convey the meaning of each word.

Actions

+   Connection

+   Expansion

+   Interruption

+   Reversal

+   Combustion

+   Separation

+   Intervention

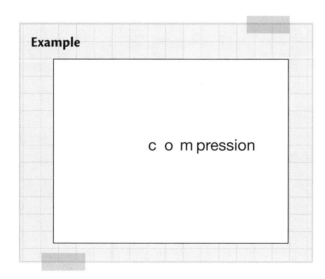

Example

c o m pression

2   Create a series of visualisation sketches that use typography as the main design element. You may also choose to use other design elements and principles to enhance the design idea. Each sketch should be an identity design concept for one or more of the following organisations:

+   'The Vintage Garden' Antique Fair

+   'High Spirits' Flying Trapeze and Gymnastics Club

+   'Digitext' website copywriters

+   'Crumbs' cakes.

3   Using a combination of type and image, create digital presentations of the following proverbs. Research the meaning of each phrase and ensure that your design concept reinforces your findings. Remember to document the source of your images.

a   'A stitch in time saves nine.'

b   'A chain is only as strong as its weakest link.'

c   'A picture paints a thousand words.'

d   'Jack of all trades, master of none.'

e   'Beauty is in the eye of the beholder.'

ISBN 9780170401791

9.3 COMPOSITION AND LAYOUT

TASKS

1   Identify and draw the grid used in the following designs (for the colour versions
     of these designs see pages 215–16):

a

Adam Lipszyc

*Winner of the Don Metcalf Scholarship from
the Walter and Eliza Hall Institute, Adam is
currently working on Dengue Fever, splitting
cells and playing with mice. He has applied to
study Medicine at Sydney University in 2016.*

**Tell me about the Science rooms at
Woodleigh?** It's a really nice looking
Science building – I'm no architect but I
think it's won a few design awards.

**Did you focus mainly on the Sciences at
school?** I über-nerded it through Physics,
Chemistry and Biology. I didn't think
there was that much of a spectrum. There
were the Sciences and Maths.

**You didn't notice the large amount of
Art subjects?** I had a really shallow view
of all the other faculties, I thought Arts
was drawing and painting. Until (sadly)
much later, I found it was so much more.
I thought engineering was working on
engines, I didn't realise there was civic,
aeronautical or the other thousands – so I
was like – I don't want to work on engines
or draw!

Sophie Coleman

Nicolas Cary

b

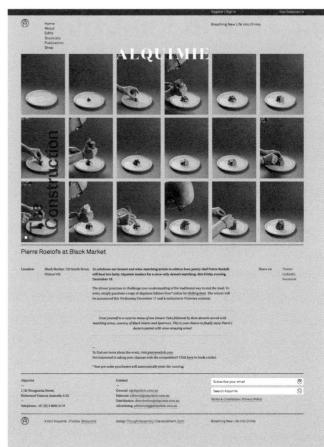

ISBN 9780170401791

c

Ryan Wheatley

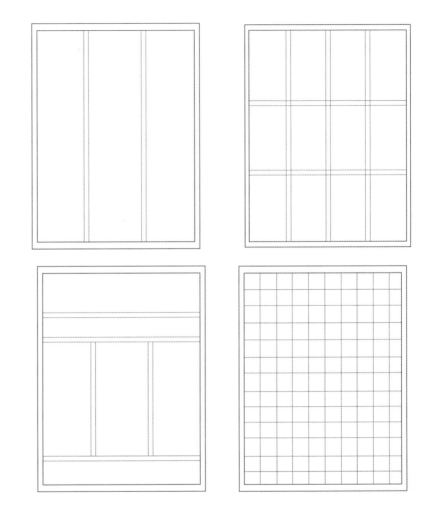

2 Using the four grids illustrated on the right, create four alternative visualisation sketches for a magazine cover all about you! Use drawing and/or collage to depict the placement of key visual information or recreate the grid in a digital form to create your design concept.

Consider the need for the following:

- title
- subtitles indicating content
- barcode
- cover image or images
- date, month, issue number.

3 Select a popular webpage, movie poster or book cover design. Identify the grid that has been applied to the design. Create an alternative layout using a grid of your own design. Establish an alternative hierarchy and apply new type and imagery while retaining the same information from the original. Present your design and the original source design side by side.

ISBN 9780170401791

PART C

ANALYSIS OF VISUAL COMMUNICATIONS

In this chapter you will focus on analysing visual communications. You will look at the manner in which the audience is targeted and evaluate whether the purpose is achieved within the context of the visual communication. You will analyse the use of design elements and design principles along with media, materials and methods. You will learn to use appropriate terminology that demonstrates an understanding of the effectiveness and impact of visual communications.

You will find that Chapter 10 in *Nelson Visual Communication Design VCE Units 1–4* will help you with this chapter.

10.1 HOW TO ANALYSE VISUAL COMMUNICATIONS

Design analysis grid

The design analysis grid enables you to collect information about visual communications. The information will assist you when describing and analysing the example in more detail.

The grid provides a framework for your understanding of visual communications. It enables you to take a close look at the content and understand why imagery, materials, elements and principles have been applied. Once you have this information, you can analyse the effect these components have on the audience and purpose of the visual communication.

The grid is shown over the next few pages in the form of 10 separate tables:

1 visual communication description
2 audience
3 purpose
4 context
5 design elements
6 design principles
7 methods
8 media
9 materials
10 other visual devices.

Key questions

The following key questions are vital to effective visual communication analysis:

+ What? What elements, principles, materials, methods and media have been used?

+ Why? Why have the elements, principles, materials, methods and media been used?

+ What effect? What effect do the elements, principles, materials, methods and media have on the visual communication and how do they assist in achieving the purpose and attracting the audience?

ISBN 9780170401791

1 Visual communication description

Identify the type of visual communication	☐ Map	☐ Explanatory diagram	☐ Statistical diagram	☐ Symbol	☐ Logo	☐ Chart	☐ Illustration	☐ 3D model
	☐ Instrumental drawing	☐ Architectural drawing	☐ Poster	☐ Packaging	☐ Signage	☐ Multimedia	☐ 2D layout	☐ Other
Describe the visual communication in detail								

2 Audience

Identify the audience	Age	Gender	Interests	Socioeconomic status	Cultural or religious factors
Describe the visual indicators that suggest the audience					

ISBN 9780170401791

3 Purpose

Identify the purpose(s) of the visual information	☐ Promote	☐ Advertise	☐ Depict	☐ Explain	☐ Teach	☐ Inform	☐ Guide
Describe how the visual information achieves this purpose							

4 Context

Identify where the visual communication is located	
Describe the impact the context has on the appearance and content of the visual communication	

5 Design elements

Indicate the design elements used	☐ Colour	☐ Form	☐ Type	☐ Line	☐ Point	☐ Shape	☐ Texture	☐ Tone
Describe the visual effect they have within the visual communication								

6 Design principles

Indicate the design principles used	☐ Balance	☐ Contrast	☐ Cropping	☐ Figure–ground	☐ Hierarchy	☐ Pattern	☐ Proportion	☐ Scale
Describe the visual effect they have within the visual communication								

ISBN 9780170401791

7 Methods

Indicate the methods used	☐ Drawing, including instrumental and freehand	☐ Printing, including printmaking techniques	☐ Photography, including digital photography and montage	☐ Collage and other illustration techniques	☐ Computer-generated content	☐ 3D process, including manufacturing processes
Describe the effect(s) the methods have on the visual communication and suggest why these methods were chosen						

8 Media

Identify the media used	☐ Pencil, pen or other drawing media	☐ Paint or other wet media	☐ Digital media
Describe the visual effect of the media and suggest why these media were chosen			

ISBN 9780170401791

9 Materials

Identify the materials	☐ Paper	☐ Card	☐ Wood	☐ Glass	☐ Clay	☐ Stone	☐ Textile	☐ Plastic	☐ Metal
Describe the visual effect of the materials used and suggest why these materials were chosen									

10 Other visual devices

Identify any other visual devices used	☐ Humour and metaphor	☐ Social comment	☐ Political comment	☐ Emotive imagery	☐ Historical or cultural references	☐ Other
Describe the impact of additional visual devices and suggest why they have been applied						

ISBN 9780170401791

TAKEAWAY

 **TASKS**

1   Create a glossary of design elements and design principles. When you analyse visual communications, it is important that you have a strong grasp of appropriate terminology and suitable analytical language at your disposal. Building a glossary – or list of words and phrases – related to design elements and design principles that you can quickly refer to can be very helpful.

   • Complete lists for all of the design elements and the design principles in your glossary. Write down as many descriptive terms as you can.

      For example: **Colour** – contrast, emotion, symbolism, vivid, subdued, vibrant, bright, energy, hierarchy, dominant, background, pattern, subtle, warm, cool, primary, secondary, eye-catching.

      For example: **Contrast** – creates visual interest; creates tension within a design through use of opposites: large–small, light–dark, cool colours–warm colours, serif–sans serif, matt–glossy, transparent–opaque.

   You may wish to do the same for materials, media and methods.

10.2 ANALYSIS THROUGH OBSERVATION

You will find that Chapter 10, pages 232–3 of *Nelson Visual Communication Design VCE Units 1–4* will help you to complete the tasks in this section.

Visual communication 1: Revolution energy drink

ISBN 9780170401791

1   Look at the poster for the energy drink (see page 216 for the colour version). The poster is for use in retail stores, cafes and food courts. The poster will also be seen at tram and bus stops.

a   Complete the table.

Describe the purpose of the poster	Explain how the poster has been designed to achieve the purpose

b   Discuss the application of shape and type to create hierarchy in the poster.

_____

_____

_____

_____

_____

_____

_____

_____

c   i   Describe the use of colour in the poster.

_____

_____

_____

_____

_____

_____

_____

_____

ii   Explain how colour has been used to assist in achieving the purpose.

_____

_____

_____

_____

_____

_____

_____

_____

_____

d   Suggest two other visual communication presentations that could be used to promote the energy drink.

i   _____

ii   _____

Visual communication 2: Poster for MYCONCIERGE service

2 Look at the large-format poster for a concierge assistance service (see page 217 for colour version). The poster is displayed only in the domestic and international terminals of airports within Australia. Small versions of the poster can be found in airline business lounges.

a Complete the following table.

Describe the target audience of the poster	Explain how the poster is designed to appeal to the target audience

b Describe the application of one design principle in the poster.

Design principle:

_____

_____

_____

_____

_____

_____

_____

_____

c Discuss how the application of this design principle is effective in attracting the target audience.

_____

_____

_____

_____

_____

_____

_____

_____

_____

_____

_____

_____

d Describe the application of one dominant design element (other than type) in the poster.

Design element:

_____

_____

_____

_____

_____

_____

_____

_____

_____

_____

_____

_____

_____

ISBN 9780170401791

e   Discuss how the application of this design element is effective in attracting the target audience.

_____

_____

_____

_____

_____

_____

f   The poster is large format. Suggest why such a scale has been used and how this influences the effectiveness of the visual communication.

_____

_____

_____

_____

_____

_____

_____

_____

STERLING CRUISES

A grand ocean voyage inspired by the glories of travels golden age.

This is the trip of a lifetime and the opportunity to view locations many can only dream of. Every moment on board the magnificent Queen of the Seas is an unforgettable experience. We invite you to become a part of an experience beyond compare.

A Sterling World Cruise.

Reserve your place in history.
Call Sterling Cruises now 13 17 65
or visit sterlingcruises.com

iStockphoto/garywg (ship)

Visual communication 3: Sterling Cruises

ISBN 9780170401791

3 Look at the front page of the brochure for Sterling Cruises (see colour version on page 217).

   a Suggest a likely context for the brochure and explain your answer.

_____

_____

_____

_____

_____

_____

_____

_____

   b Describe the target audience for the brochure and support your answer with visual evidence.

_____

_____

_____

_____

_____

_____

_____

_____

_____

_____

_____

   c What is the purpose of the brochure? Support your answer.

_____

_____

_____

_____

_____

_____

_____

_____

_____

   d Describe two methods used in the production of the brochure.

    i _____

_____

_____

_____

_____

_____

    ii _____

_____

_____

_____

_____

_____

e Discuss the effectiveness of one design principle in the brochure design.

_____

_____

_____

_____

_____

_____

_____

_____

_____

_____

_____

f Discuss the effectiveness of one design element in the brochure design.

_____

_____

_____

_____

_____

_____

_____

_____

_____

_____

_____

g Describe a way the designer of the Sterling Cruises brochure might have used digital media during the design process.

_____

_____

_____

_____

_____

_____

_____

_____

_____

_____

_____

_____

_____

_____

_____

_____

_____

_____

_____

_____

_____

_____

_____

ISBN 9780170401791

Chapter 11
DESIGN FIELDS

There are three main design fields identified in VCE Visual Communication Design: industrial design, environmental design and communication design. Each design field has its own language and traditions, origins and influences that distinguish it from other professional fields. Chapter 11 of *Nelson Visual Communication Design VCE Units 1–4* will help you with this chapter.

11.1 INDUSTRIAL DESIGN

Industrial designers create consumer or industrial products both large and small, including but not limited to motor vehicles, consumer electronics, lighting, furniture, medical equipment, toys, recreational products, industrial machinery and water craft. Most industrial designers strive to create products that are sustainable, efficient and effective by using innovative technologies and materials, the effective use of elements and principles and appealing aesthetics.

 TASKS

1 Select one of the designs illustrated and investigate the following (for colour versions see pages 217–18):

+ Design: _____

+ Who is the designer? _____

+ What nationality is the designer? _____

+ In what year was it designed? _____

+ What materials are used? _____

_____

Mark Wilken

Mark Wilken

Mark Wilken

ISBN 9780170401791

2   Collect three other designs by the same designer. Sketch or paste the image into the space provided. Annotate with the title of the design (if available), date and source of the image. Make sure you appropriately cite your source.

3   Name two other designers who were (or are) actively designing during the same period as your designer:

Designer 1: _____

Designer 2: _____

4   Select a *visual* example of each designer's work. Sketch or paste the image into the space provided. Annotate with their name, the title of the work and the date it was made. Make sure you appropriately cite your source.

5   Write a short (250-word) biography of your designer including their background, education, influences and key designs for which they are known.

_____
_____
_____
_____
_____
_____
_____
_____
_____
_____
_____
_____
_____
_____
_____
_____
_____
_____
_____
_____
_____
_____
_____
_____
_____
_____
_____
_____
_____

ISBN 9780170401791

Industrial design: Furniture

6 Select one historical and one contemporary example of chair design; for example, the Red and Blue Chair by Gerrit Rietveld (historical) and the Embryo Chair by Marc Newson (contemporary). Analyse your examples by completing the table.

	Historical example: _____ Year: _____ Designer (if known): _____	Contemporary example: _____ Year: _____ Designer (if known): _____
Write your own description of each chair design.		
Describe the use of design elements, such as: + colour + texture + form.		
Describe the use of design principles, such as: + proportion + balance + scale.		
Describe construction methods and use of materials, such as: + wood + fabric + paint.		
Suggest possible influences and inspirations that may have affected the appearance of the two chair designs.		

ISBN 9780170401791

Industrial design: Computers

7   Select one historical and one contemporary example of design for technology; for example, the Apple IIe from 1983 and the current iMac. Analyse your examples by completing the table.

	Historical example: _____ Year: _____ Designer (if known): _____	Contemporary example: _____ Year: _____ Designer (if known): _____
In your own words, describe the appearance of the different product designs.		
Describe the use of design elements, such as: + colour + shape + form.		
Describe the use of design principles, such as: + proportion + balance + scale.		
Describe construction methods and use of media and materials, such as: + plastic + glass + metal.		
Suggest the influences that may have affected the design of each product, such as: + materials + ergonomics + fashion.		

ISBN 9780170401791

11.2 ENVIRONMENTAL DESIGN

Designers working in the environmental design area include architects, interior designers and landscape architects. All work with the creation of functional and aesthetically appealing spaces. Their expertise ranges from domestic buildings to skyscrapers, from hospital interiors to artists' studios, from small urban landscapes to botanical gardens. All designers understand the design, construction and evaluation methods specific to their area of professional expertise. However, many professionals often work collaboratively across the design area, particularly on large-scale projects.

 TASKS

1   Select one of the two designs illustrated and investigate the following:

+   Design: _____

+   Who is the designer? _____

+   What nationality is the designer? _____

+   In what year was it designed? _____

+   What materials are used? _____

_____

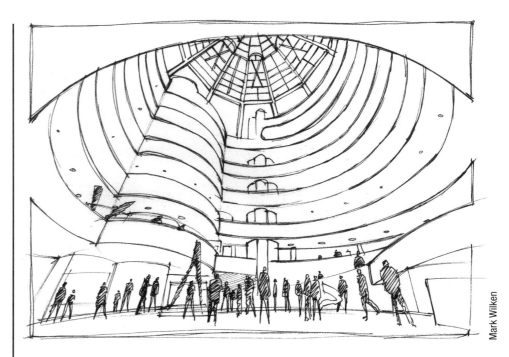

Mark Wilken

Mark Wilken

ISBN 9780170401791

2   Now that you have discovered who your mystery designer is …

a   Collect three other designs by the same designer. Annotate with the title of the design (if available), date and source of the image.

Design	Title	Date	Source
Insert an image of the design or draw your own sketch in the style illustrated on page 141.			*Copyright guidelines require that you state the URL and date accessed or book title, author and publisher.*

b Name two other designers who were/are actively designing during the same period as your designer:

Designer 1: _____

Designer 2: _____

c Select a visual example of each designer's work and provide their name, the title of the work and the date it was made.

Design	Title	Date	Source
Insert an image of the design or draw your own sketch.			Copyright guidelines require that you state the URL and date accessed or book title, author and publisher.

d Write a short (250-word) biography of your designer including their background, education, influences and key designs for which they are known.

_____

_____

_____

_____

_____

_____

_____

_____

_____

_____

_____

_____

_____

_____

_____

_____

_____

_____

_____

_____

_____

_____

Environmental design: Architecture

3 Select one contemporary and one historical example of architecture; for example, the Glass House (1949) by Philip Johnson and the Tattoo House (2007) by Andrew Maynard. Analyse your examples by completing the table.

	Contemporary example: _____ Year: _____ Designer (if known): _____	Historical example: _____ Year: _____ Designer (if known): _____
Describe the architectural examples in your own words.		
Describe the use of design elements, such as: + texture + colour + form.		
Describe the use of design principles, such as: + proportion + hierarchy + scale.		
Describe the use of materials and construction methods in each design.		
Describe the influences of each architect and reflect on their representation in the examples.		

ISBN 9780170401791

11.3 COMMUNICATION DESIGN

Communication designers work with type and image to create a wide range of graphical products in print and digital media. Common projects for graphic designers include logos and corporate branding, packaging, posters, signage/ wayfinding systems, publication design, motion graphics, web design and interactive multimedia.

Alamy Stock Photo/Mario Mitsis

Shutterstock.com/Savvapanf Photo

TASKS

1   Select one of the designs on this page and investigate the following (for colour versions see page 218):

+ Design: _____

+ Who is the designer? _____

+ What nationality is the designer?_____

+ In what year was it designed? _____

+ What materials are used? _____

_____

ISBN 9780170401791

2   Now that you have discovered who your mystery designer is . . .

   a   Collect three other designs by the same designer. Annotate with the title of the design (if available), date and source of the image.

Design	Title	Date	Source
Insert an image of the design or draw your own sketch in the style illustrated on page 137.			*Copyright guidelines require that you state the URL and date accessed or book title, author and publisher.*

b Name two other designers that were/are actively designing during the same period as your designer:

Designer 1: _____

Designer 2: _____

c Select a visual example of each designer's work and provide their name, the title of the work and the date it was made.

Design and designer's name	Title	Date	Source
Insert an image of the design or draw your own sketch.			Copyright guidelines require that you state the URL and date accessed or book title, author and publisher.

d Write a short (250-word) biography of your designer including their background, education, influences and key designs for which they are known.

_____

_____

_____

_____

_____

_____

_____

_____

_____

_____

_____

_____

_____

_____

_____

_____

_____

Communication design

3   Select one contemporary and one historical example of communication design; for example, the movie poster for *Vertigo* designed by Saul Bass (1958) and a poster for a new release film. Analyse your examples by completing the table.

	Contemporary example: _____ Year: _____ Designer (if known): _____	Historical example: _____ Year: _____ Designer (if known): _____
Describe the designs in your own words.		
Describe the use of design elements, such as: + colour + shape + type.		
Describe the use of design principles, such as: + balance + unity + space.		
Describe the use of media and illustration methods in each design.		
Describe what may have inspired or influenced the designer.		

ISBN 9780170401791

Chapter 12
INTELLECTUAL PROPERTY

Regulations apply to designs created and sold in Australia and include copyright, intellectual property rights, safety regulations and standards that designers are required to adhere to. You will find that chapter 13 of *Nelson Visual Communication Design VCE Units 1–4* will help you with the tasks in this chapter.

 TASKS

Case studies

1   Read the following brief case studies. Identify the legal responsibilities and related actions that designers may need to address. Use chapter 13 of *Nelson Visual Communication Design VCE Units 1–4* to assist in writing your responses.

Case	Legal requirements	What action might be taken to meet the legal requirements?
A student studying VCE Visual Communication Design wishes to use the well-known logo of a sportswear brand as part of a school-based design task. The logo will only be seen on the student's work, which is to be displayed at the school's annual Art and Design exhibition.		
An interior designer wishes to use the image of a recognisable and popular musician as part of a mural at the headquarters of the Australian Music Foundation. The mural design will also feature imagery of the album artwork and stills from a music video.		

ISBN 9780170401791

Case	Legal requirements	What action might be taken to meet the legal requirements?
A landscape architect has been employed to design an innovative children's play area in a public park. The park is used by children and families from the local area. The park is also an off-lead dog area, and on weekends is used for junior-level soccer matches.		
A team of industrial designers are working with engineers on the design of a new coffee machine for domestic use. It will feature rapid heat settings and an integrated milk warmer/frother.		
An architectural firm has been contracted to design a new transport hub at the main city airport. The hub will accommodate buses, a light rail system and valet car parking. It will offer sheltered areas for waiting passengers as well as café and retail areas.		
A communication design studio is designing the promotional materials, including posters, online content and tickets, for an upcoming exhibition of works by renowned Australian artist Howard Arkley. Images of Arkley's work will feature prominently in all advertising collateral.		
A school plans to present a public performance of the musical *Grease*. They will promote the event using billboards on the street and on posters in local businesses. T-shirts featuring the logo will be for sale on the opening night.		

ISBN 9780170401791

Intellectual property

2   Using the table provided on page 287 of *Nelson Visual Communication Design VCE Units 1–4*, research and identify an example of each of the following intellectual property rights.

Type of IP right	Example of protected intellectual property
Patents	
Trade marks	
Designs	
Copyright	

ISBN 9780170401791

Creative commons

3 Identify the meaning of the following creative commons licences. Go online and find an example of where each licence has been used and add the name of the site, type of content and URL.

Creative commons licence	Meaning	Example of use

Part D

Chapter 13

WRITING THE DESIGN BRIEF

13.1 THE DESIGN BRIEF

The design brief is the starting point from which the final two visual communication presentations are produced. Your design brief should outline the requirements and expectations of your client, including the design requirements, purpose, context, design constraints and intended target audience.

You will find that Chapters 8 and 10 of *Nelson Visual Communication Design VCE Units 1–4* will help you with this section.

Understanding the client and design task

 **TASKS**

On receiving a design brief, it is important to 'unpack' the communication need or needs that you will be required to solve. Clarify the brief with a 'return brief' so that you, as the designer, have a clear understanding of the client's expectations.

1 Client name: _____

Explanation of reason for design. What is the communication need?

_____

_____

_____

_____

_____

_____

_____

_____

2 Client's background and design history. List previous or existing designs.

_____

_____

_____

_____

_____

_____

_____

_____

_____

_____

3 List any non-negotiable inclusions, such as logo, corporate colours, branding and so on.

_____

_____

_____

_____

_____

_____

_____

4 List any general client preferences for elements, such as colour, appearance, type and so on.

_____

_____

_____

_____

_____

_____

_____

5 What are the client's expectations of deliverables (final presentations)?

_____

_____

_____

_____

_____

_____

_____

_____

_____

_____

_____

_____

ISBN 9780170401791

Understanding the audience

TASKS

See Chapter 5 for more information and tasks related to understanding the audience.

1   Age range of target audience _____

2   Socioeconomic status   ☐ High   ☐ Medium   ☐ Low

3   Disposable income (if relevant) ☐ High   ☐ Medium   ☐ Low

4   Interests and tastes

_____

_____

_____

_____

_____

Brands: which?	Music: what?	Magazines and TV: what?	Social life: what and where?	Shopping: what and where?	Hobbies and sports: what?

ISBN 9780170401791

5 Where does your audience spend most of their time?

☐ School/university ☐ Indoors
☐ Work ☐ Alone
☐ Home ☐ In company
☐ Outdoors ☐ In an office

6 Which key words describe the target audience?

☐ Contemporary ☐ Carefree
☐ Fashion-conscious ☐ Cheap
☐ Adventurous ☐ Expensive
☐ Cutting-edge ☐ Natural
☐ Alternative ☐ Extravagant
☐ Ethical ☐ Stylish
☐ Conservative ☐ Safe or cautious
☐ Environmentally focused ☐ Family-focused
☐ Politically minded ☐ Youthful
☐ Social ☐ Technology-savvy

7 Where is the audience located?

_____

8 What impact might their location have on the audience?

_____
_____
_____
_____

9 List any other words that describe the audience:

_____
_____
_____
_____
_____
_____

10 What is important to the audience?

☐ Social and political issues ☐ High quality
☐ The environment ☐ Low price
☐ Technology ☐ Contemporary design
☐ Social networking ☐ Traditional design
☐ Saving money ☐ Innovation
☐ Spending money ☐ Tradition
☐ Luxury ☐ Natural things
☐ Security ☐ Manufactured things

11 List the images that come to mind when you reflect on the target audience.

_____   _____
_____   _____
_____   _____
_____   _____

ISBN 9780170401791

12 Describe the aesthetic preferences of the audience by completing the table.

Colours	
Shapes and forms	
Patterns and decoration	
Materials and textures	

Understanding the constraints of the design brief

A design problem will often come with constraints, whether these are related to cost, time or access to materials, all will have an impact on the design process.

 TASKS

1 Read the following design problems and identify possible constraints that may affect the design process. Suggest how the constraints may influence the decision making of a designer.

Example

Design area	Design problem	Likely constraints	Possible decisions that may be made to address constraints
Commuciation design	To create a simple, illustrated first-aid booklet to be distributed to all junior primary school students in Queensland. The booklet will feature instructions about basic treatments for scratches, stings and minor injuries.	*Cost – due to high number of booklets to be produced.*	*Consideration may need to be given to paper quality, full-colour content and printing costs.*

ISBN 9780170401791

Design field	Design brief	Likely constraints	Possible decisions that may be made to address constraints
Industrial design	The design of an underwater camera casing for action cameras. The casing should suit a range of designs on the market and allow for maximum functionality when underwater.		
Environmental design – Architecture	The design of a new domestic dwelling in a cyclone-prone region of northern Australia.		
Environmental design – Interior design	The design of a new gymnasium and wellness centre to service a large urban population. The users will be of mixed age and mobility.		
Environmental design – Landscape architecture	The design of an urban landscape including a park and playground area that is open to the public. It will be opened at the same time as a soon-to-be-completed civic centre.		
Communication design	The design of a web page that is also functional on a range of mobile devices. The website is for the promotion of an upcoming comic convention.		

2 Identify the constraints contained within the following design brief. Circle or highlight each of the constraints.

Circa 1602 Trading Co. is a boutique specialist spice, tea and herb trading company. Circa 1602 is a boutique trading company and is seeking a sophisticated, high-end identity design that reflects its ethos. The company is inspired by the Dutch East India Trading Company, which was founded in 1602 when the States-General of the Netherlands granted it a 21-year monopoly to carry out colonial activities in Asia; the name of the company, 'Circa 1602', stems from this significant historical date.

The company is based in Brisbane and predominately markets its products to specialist stockists such as providores and delicatessens, as well as online customers. Circa 1602 prides themselves on providing high-quality spices, teas and herbs; and because of its quality ethos and artisan products the company has strived to serve a high-end market; targeting a customer base with high disposable incomes. Circa 1602 Trading Co. requires the design of its corporate identity, its logo and accompanying stationary: business card, letterheads, etc. The main purpose of the identity design is to promote and advertise Circa 1602 Trading Co. to gourmet and boutique stockists, restaurants and chefs, as well as individual customers of local boutique providores and gourmet produce stores.

The target audience for the corporate identity of Circa 1602 is potential stockists: primarily owners of boutique providores and delicatessens, restaurant owners and chefs, and specialty hospitality industry suppliers; as well as customers of these stockists and suppliers.

The corporate identity design must be cost effective, as the budget is moderate due to general business set-up costs. The identity should feature the logo for Circa 1602, and will be applied to company signage, merchandise, business cards, web pages, shipping and freight packaging, etc. Therefore, the logo must be effective in both colour and/or black and white, and able to be applied to a range of visual carriers, work in various sizes/proportions, and be representative of Circa 1602's quality ethos and influences.

ISBN 9780170401791

Your design brief

Defining the design problem and writing a design brief.

 **TASK**

Fill in the design brief template.

Design brief template

Describe the communication need; what does the client require you to design?						
Client name (if relevant)						
About the client (summary)						
What is the main purpose or purposes of the design?						
Audience information	Age:	Gender:	Interests:	Socioeconomic status:	Cultural or religious factors:	Location:
Constraints	Time factors	Cost factors	Location	Materials	Technologies	Other

ISBN 9780170401791

Context *Where (location) or when (time) will the design be used/seen?*		
Considerations	**Design elements and design principles** Are there any *required* design elements and design principles? If so, describe.	
	Media, methods and materials Have any media, methods or materials been specified by the client? If so, describe.	
	Legal and ethical responsibilities Are there relevant legal responsibilities (copyright and IP, standards, etc.) that may impact the design process? If so, describe.	
	Sustainability Are there any sustainability factors that should be taken into consideration? If so, describe.	
Deliverables *What are the final presentations to be delivered?*		

Chapter 14
RESEARCH

Researching your design brief can be daunting. Resist the temptation to fill pages with images; be selective in gathering your research and find images and ideas that inspire your own concepts.

What can you research?

Research related to the client	Research of the target audience	Research of similar graphical representations	Research for inspiration and ideas
Background	Tastes/preferences	Current examples	Ideas to help generate concepts for your design (these may be unrelated to your design but may offer creative inspiration, e.g. the form of a random object)
Previous designs they have used	Interests	Historical examples	
Who are their competitors?	What products/designs/spaces do they use?		

Quality of research

 **TASKS**

1   Identify whether the following examples of research are qualitative or quantitative (tick the correct column):

Example

Research method	Qualitative	Quantitative
Analysis of statistical information from the Australian Bureau of Statistics derived from the Census.		✓

Research method	Qualitative	Quantitative
Precise measurements taken at a building site to establish the position of the new domestic dwelling.		
Interview with the typical user of new wireless headphones.		
Focus group discussion of poster designs for a travel destination.		
Analysis of feature article from a popular blog.		
Collation of data that states the number of people entering and exiting a public building at a given time of day.		

2   Identify whether the following examples of research originated from **primary source/s** or **secondary source/s** (tick the correct column):

Research method	Primary source	Secondary source
Industrial designer interviews the user of a convertible motor vehicle about their likes and dislikes.		
Interior designers discuss specific aspects of a design brief with the client.		
Collection of purchased stock images used to create a 'look and feel' board for a landscape architecture brief.		
Collection of quotes and comments from an online newspaper article about a newly commissioned corporate logo design.		
Architect takes photographs and video of the environment and topography, on a site visit.		

Research plan for a handbag

Example

Initial ideas	Type of research	What am I looking for?	How will I undertake the research?
Handbag design	Interviews with current users	Information about what works and what doesn't.	Email questions to six friends
	Images of bags	Inspiration for forms and details such as handles, closures, etc.	Images from the Internet and fashion magazines
	General inspirational images	Inspiration for patterns and textures; take photos of textures around me. Use my own photographs of forms and patterns that I think will suit my audience.	Visit local buildings and markets to take photos

ISBN 9780170401791

My research plan

Initial ideas	Type of research	What am I looking for?	How will I undertake the research?

ISBN 9780170401791

Once you have collected your research, another method of organisation is to use a 'design scale'. Place your research within the scale and identify where you would like to position your own design. This method is also helpful generating concepts.

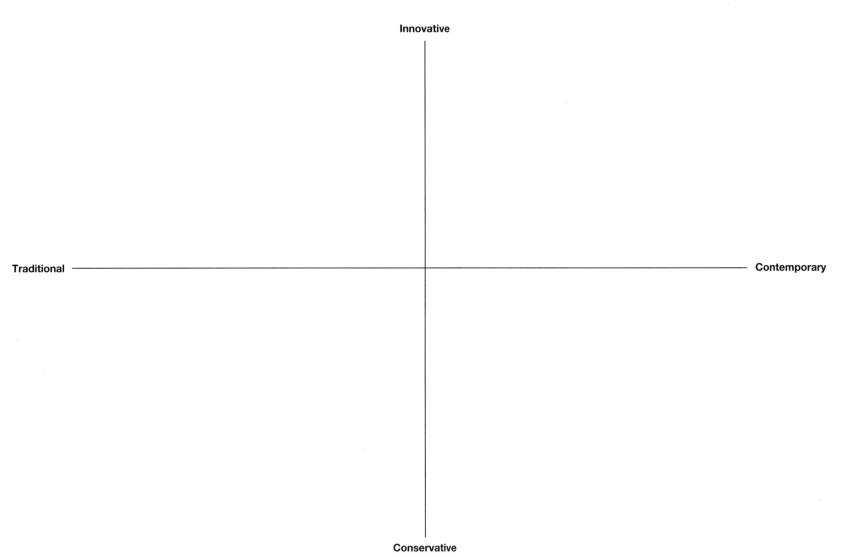

Innovative

Traditional ——————————————————————— **Contemporary**

Conservative

ISBN 9780170401791

Chapter 15

GENERATION, DEVELOPMENT & REFINEMENT

15.1 GENERATING

There are many ways to begin generating concepts in your folio.

Remember, the effective use of media, materials, methods, design elements and design principles should be evident, even at this stage.

Use Chapter 1 of this workbook and Chapters 8 and 9 in *Nelson Visual Communication Design VCE Units 1–4* to access design thinking tools to help you to generate ideas.

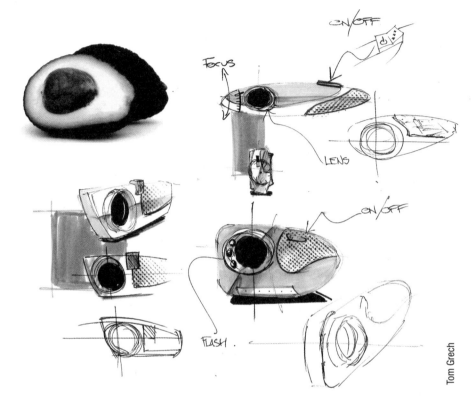

Tom Grech

This student used an image of an avocado to inspire ideas for a digital camera design.

ISBN 9780170401791

 **TASKS**

1 Start by asking questions and use drawing, rather than words, as a response.

Where will the design be located or used?	Who uses or sees the design?	What does the design need to do or achieve?	What features might the design include?	How will the design be distinctive?

ISBN 9780170401791

2 The grid below lists methods of generating concepts. You may choose to use this grid to identify starting points and to set expectations of the work required at this early stage of the design process. Use the relevant chapters, as indicated in both this workbook and in *Nelson Visual Communication Design VCE Units 1–4*.

You may elect to use one or several starting points to begin generating ideas.

Starting point	Details	Notes and ideas
Inspiration images	The placement of images on the generation page to inspire ideas. + Drawings link initially to the inspiration image and then diversify as they evolve.	
Look and feel	Collection of images, texture and colour swatches, patterns, etc., that identifies the 'mood' or 'feel' of the proposed design. The 'look and feel' is designed for inspirational purposes; elements may find their way into the design concepts.	
Audience collage	A collage of images, textures, colours and type related specifically to the target audience. Assists in identifying preferences of the audience and may suggest: + colour palette + shapes and forms + pattern, texture and decorative elements + materials.	

ISBN 9780170401791

Starting point	Details	Notes and ideas
Random words	Similar to inspiration images (page 167) but using words rather than visuals. Particularly effective in generating ideas for product design and in logo generation. Sample words might be 'bold', 'gentle', 'active', 'elegant' and so on.	
Photography	Sketches and ideas can be generated from photographs taken by you. The original photo(s) should be placed on the generation page to indicate the source.	
3D	If you initially find drawing difficult, a concept can be constructed from paper or card. The concept can then be drawn from direct observation or you might photograph the concepts and annotate. This technique is ideal if you like to work with your hands and appreciate visualising in 3D.	

Look and feel

TASKS

Used in all design areas, 'look and feel' presentations (also known as mood boards) can assist in defining the likely visual and aesthetic direction of a design. These can be essential in communicating design ideas to a client in the early stages of the design process.

1   Observe the look and feel boards pictured on pages 176–7 of *Nelson Visual Communication Design VCE Units* 1–4. In the space below, list words that describe the likely 'look' and 'feel' that the designers were aiming for.

a

_____

_____

_____

_____

_____

_____

b

_____

_____

_____

_____

_____

_____

2   On a separate page using either cut and paste technique (or by creating your own document in software such as Photoshop), create a look and feel presentation that captures the one of the following descriptions.

a   Interior design: sophisticated, calm, luxurious, indulgent and comfortable.

b   Industrial design: sleek, contemporary, useful, aspirational, ergonomic and vibrant.

c   Landscape architecture: meditative, peaceful, quiet, natural, organic and accessible.

d   Architecture: solid, innovative, patterned, sustainable, welcoming, engaging and tactile.

e   Communication design: stimulating, patterned, urban, youthful, rebellious and retro.

ISBN 9780170401791

Sketching ideas

 TASKS

Sketching is an essential part of generating design ideas. Quick and easily executed, hand-drawn sketches offer immediate creative input to a design problem. They are an important part of the process. Practice is the key to building skills in sketching; the more you draw the better your skills will become.

Observational sketches are an established method of building your skills and confidence in sketching. Using objects that are around you, look closely at their features, proportions and textures. Drawing the familiar will see your skills build very quickly.

1 Select a simple object that is readily accessible, such as a pencil sharpener, your shoe, wristwatch or eraser. In the grid provided, draw each of the views as described. Use a loose grip on your drawing medium (pencil, fineliner, etc.) and don't overwork your lines.

Top view	Three-quarter view	Zoom view
Draw the object from above.	Draw the object from a corner or edge.	Draw a close-up detail of the object.

Outline view	Rendered view	Contextual view
Draw the object using line only, no tone or shadow.	Draw the object using tone only – no outlines.	Draw the object in its environment, e.g. on the table with other objects, on your foot, on a wrist, etc.

ISBN 9780170401791

2 Drawing from your imagination can be very challenging and when faced with a blank page, some students find the process intimidating. Key to remember is that ideas are just that – ideas. You should only use a filter on your concepts as you get to the latter stages of the design process. Avoid filtering ideas at the early stages – you may find that there are many great ideas that come from random beginnings.

Pictured on this page and the next are three common objects. By drawing, convert each object into a product with a very different function.

Concept 1: a blender. Using drawing, change it into a robot.

Concept 2: A golf cart. Using drawing, convert it into a flying machine.

Shutterstock.com/ProstoSvet

iStock.com/venakr

ISBN 9780170401791

Concept 3: Remote control. Using drawing, convert it into a backpack.

Shutterstock.com/Carolina K. Smith MD

Annotation

TASKS

Annotation provides written reflection on your design ideas; it involves thinking about your thinking. Good annotation is reflective, succinct and relevant. It conveys analysis and suggests possible directions for further development.

These key concept questions will help you to make effective annotations:

- Descriptive: What were you doing?
- Predictive: Where might the idea lead?
- Reflective: Is it a good idea?
  - Does it fulfil the design brief?
  - Does it meet the purpose?
  - Does it appeal to the target audience?

Example

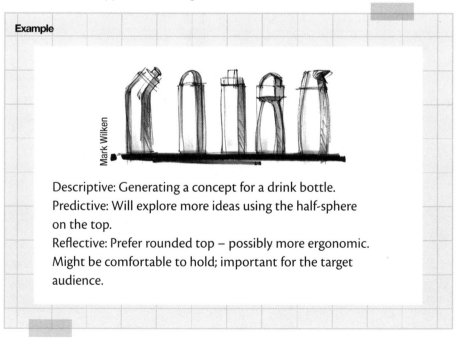

Mark Wilken

Descriptive: Generating a concept for a drink bottle.
Predictive: Will explore more ideas using the half-sphere on the top.
Reflective: Prefer rounded top – possibly more ergonomic. Might be comfortable to hold; important for the target audience.

ISBN 9780170401791

Descriptive: _____

_____

Predictive: _____

_____

Reflective: _____

_____

_____

_____

_____

Descriptive: _____

_____

Predictive: _____

_____

Reflective: _____

_____

_____

_____

_____

ISBN 9780170401791

Selecting preferred concepts

TASKS

When you develop ideas, you will trial and experiment with a range of materials, technologies and graphical representations. The use of design elements and principles should be imaginative, appropriate and effective. The idea or ideas that you choose to develop should best meet a solution for the original design problem.

Evaluating design alternatives

You may choose to draw or use text to identify your preferred concept/s.

1   Based on the strengths, weaknesses and relationship to the original design problem, list your preferred concepts in order of preference:

a _____

b _____

c _____

d _____

2   Based on the strengths, weaknesses and relationship to the original design problem, sketch and annotate your preferred concepts in order of preference.

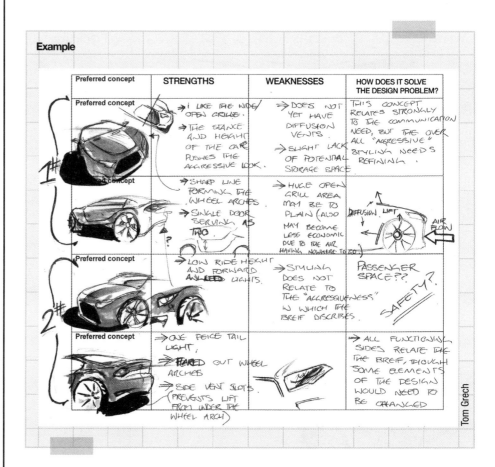

Preferred concept	Strengths	Weaknesses	How does it address the communication need?
Concept 1			
Concept 2			
Concept 3			
Concept 4			

ISBN 9780170401791

15.2 DEVELOPING

The elements and principles of design are integral to creating effective graphical products. Experimentation with a range of elements and principles can help to develop and refine your design ideas. See Chapter 2 in this workbook and Chapter 5 in *Nelson Visual Communication Design VCE Units 1–4* to assist your use of these important aspects of your design.

 ## Use of design elements

Design elements	Where have I used it?	How have I experimented/trialled/tested?	Possible further uses/experimentation?
☐ Colour			
☐ Form			
☐ Line			
☐ Point			
☐ Shape			
☐ Texture			
☐ Tone			
☐ Type			

ISBN 9780170401791

Use of design principles

Design principles	Where have I used it?	How have I experimented/trialled/tested?	Possible further uses/experimentation?
☐ Balance			
☐ Contrast			
☐ Cropping			
☐ Figure–ground			
☐ Hierarchy			
☐ Pattern			
☐ Proportion			
☐ Scale			

ISBN 9780170401791

Use of methods

This table enables you to reflect upon your use of drawing methods.

Method	Where and how have I used it?	Effective?	Needs improvement?
☐ Drawing: observational		*	*
☐ Drawing: visualisation		*	*
☐ Drawing: presentation		*	*
☐ Collage		*	*
2D: ☐ Orthogonal drawing * Packaging nets		*	*
3D: paraline ☐ Isometric ☐ Planometric		*	*

ISBN 9780170401791

Method	Where and how have I used it?	Effective?	Needs improvement?
3D: perspective ☐ One-point perspective ☐ Two-point perspective		*	*
Digital methods ☐ Vector-based ☐ Raster-based		*	*
Photography ☐ Analogue ☐ Digital		*	*
Printing: manual ☐ Monoprint ☐ Relief ☐ Intaglio ☐ Silk screen		*	*
Printing: digital ☐ Offset ☐ Laser ☐ Inkjet		*	*
3D processes ☐ Construction ☐ Modelling ☐ Digital		*	*

ISBN 9780170401791

Experimenting with materials and media

Use these tables to ensure that you have covered a range of media and materials.

Use of materials

Materials	Where is it used or represented?	Possible further uses and experimentation or representation?
☐ Paper		
☐ Card		
☐ Stone		
☐ Glass		
☐ Metal		
☐ Clay		
☐ Plastic		
☐ Textile		
☐ Screen		

ISBN 9780170401791

Use of media

Media	Where is it used or represented?	Possible further applications and experimentation?
☐ Pencil		
☐ Paint		
☐ Ink		
☐ Marker		
☐ Pastel/crayon/charcoal		
☐ Watercolour		
☐ Gouache		
☐ Film		
☐ Digital media		
☐ 3D construction materials		

ISBN 9780170401791

15.3 REFINING IDEAS

At the refinement stage of the design process, it is important for you to select the most appropriate design concept. Completing the table will help you to decide on the most suitable concept and see what else you should do to it.

Your folio should show refinement of your preferred concept with clear annotations and use of appropriate media, materials, methods, design elements and design principles.

Design brief: requirements	Preferred concept: relationship to the design brief	Refinement required to meet the design brief
☐ Purpose		
☐ Context		
☐ Audience		
☐ Constraints		
☐ Functional requirements		
☐ Aesthetic requirements		
☐ Other requirements		

ISBN 9780170401791

Chapter 16
EVALUATION & PRODUCTION

16.1 EVALUATION

 When you evaluate your visual communication, you have the opportunity to ensure that the final design meets the design brief. Use the evaluation sheet to ensure that all criteria outlined in the brief are covered (see the example).

Example

Evaluation sheet

Design brief	Description	Evaluation
Audience: young females, aged 16–18, interested in shopping, socialising and fashion.	Use of a fashionable colour palette in the visual communication is appealing to the target audience.	Market testing: showed colour and texture options to members of the target audience. 100% approval.
	Type: use of a sans serif typeface is effective in identifying key information for the audience.	Observation: Need to include a serif typeface as well to provide contrast in the design. Contrast is more likely to attract the target audience.

Design brief	Description	Evaluation

ISBN 9780170401791

16.2 FINAL PRESENTATION

Final presentation format

Identify the most appropriate final presentation to meet the requirements of the design brief. The table shows the different elements that your final presentation may incorporate.

Final presentation	Appropriateness to the design brief	Description of possible application to meet the brief
Logo		
Signage		
Flyer		
Brochure		
Poster		
Billboard		
Postcard		
Advertisement		
Map		

Final presentation	Appropriateness to the design brief	Description of possible application to meet the brief
Diagram		
Symbol/icon		
Illustration		
Book/magazine cover/layout		
Web application		
Exhibition screen display		
Film credit sequence		
3D model		
Packaging		
Point-of-purchase display		
Architectural drawing		
Finished drawings for a product design		

ISBN 9780170401791

Final presentation plans

Planning your final visual communication presentations is very important given the tight timeline for Unit 4. Ensure that you have the appropriate materials and technology to execute your final presentation in an appropriate time frame.

Final presentation 1	Materials required	Media required	Digital methods and media requirements	Scale/size	Timeline (no. of days/lessons)

ISBN 9780170401791

Final presentation 2	Materials required	Media required	Digital methods/media requirements	Scale/size	Timeline (no. of days/lessons)

ISBN 9780170401791

16.3 DESIGN PROCESS CHECKLIST

Design brief

- ☐ Client
- ☐ Audience
- ☐ Description of the communication need
- ☐ Purpose
- ☐ Context
- ☐ Relevant constraints
- ☐ Final presentation formats
- ☐ (if required) Is your brief signed and dated?

Design process

Does your book show a clear design process?
Does your book tell a visual story of the design process?

Sources

Have you acknowledged sources of visuals that are not your own? It is important to respect copyright and attribute all sources of non-original materials. Annotate to identify origins.

Annotation

Your research and your drawings/development work should be clearly annotated. Describe why you have collected/used and/or how you have developed/changed/designed. Your annotation is:

- ☐ Descriptive
- ☐ Analytical
- ☐ Predictive
- ☐ Reflective

Drawing

- ☐ Observation
- ☐ Visualisation
- ☐ Presentation

Elements and principles

Design elements:

- ☐ Colour
- ☐ Form
- ☐ Line
- ☐ Point
- ☐ Shape
- ☐ Texture
- ☐ Tone
- ☐ Type

Design principles:

- ☐ Balance
- ☐ Contrast
- ☐ Cropping
- ☐ Figure-ground
- ☐ Hierarchy
- ☐ Pattern
- ☐ Proportion
- ☐ Scale

Media, materials and methods

Methods:

- ☐ Photography
- ☐ Drawing

Representations of 2D and/or 3D drawing:

- ☐ One-point perspective
- ☐ Two-point perspective
- ☐ Isometric
- ☐ Planometric
- ☐ Orthogonal

Packaging net:

- ☐ Digital methods
- ☐ 3D construction
- ☐ Printing/photocopier

Materials:

- ☐ Paper
- ☐ Card
- ☐ Plastic
- ☐ Textile
- ☐ Wood/stone/ metal/glass
- ☐ Screen

Media:

- ☐ Pencil
- ☐ Ink
- ☐ Marker
- ☐ Pastel
- ☐ Crayon
- ☐ Charcoal
- ☐ Acrylic paint
- ☐ Watercolour
- ☐ Gouache
- ☐ Dye
- ☐ Toner
- ☐ Film
- ☐ Digital applications
- ☐ Vector-based programs (e.g. Illustrator)
- ☐ Raster-based programs (e.g. Photoshop)

Final presentations

Do your presentations show imagination and creativity? Are they presented with technical skill? Is there a clear relationship between the final presentations and the original design brief?

ISBN 9780170401791

APPENDIX I: DRAWING GRIDS

These grids may assist you in completing your instrumental drawings.

Isometric grid

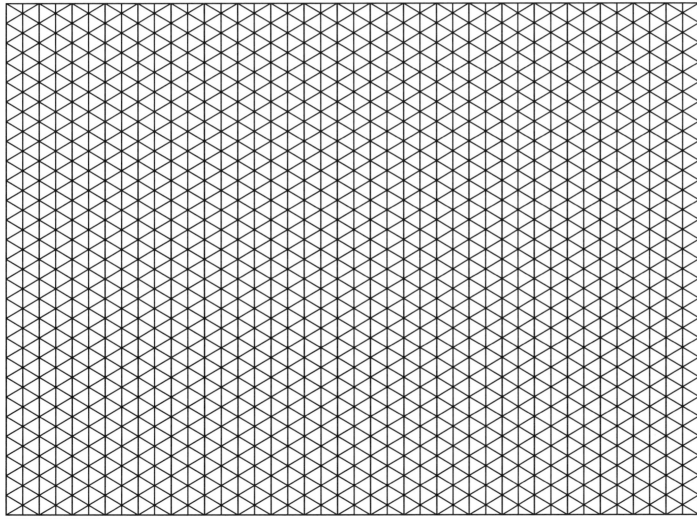

ISBN 9780170401791

Planometric grid

ISBN 9780170401791

3 mm grid

ISBN 9780170401791

5 mm grid

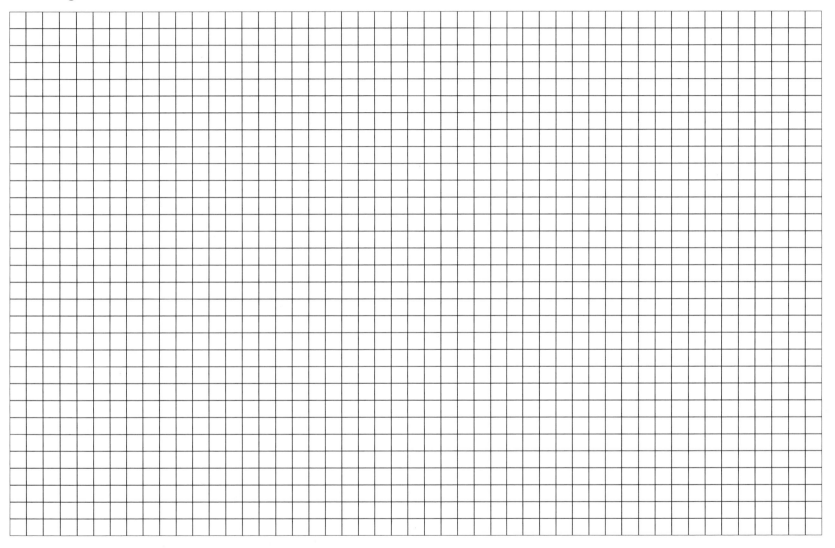

ISBN 9780170401791

Horizontal grid (lined paper for lettering)

ISBN 9780170401791

APPENDIX II: SELECTED WORKBOOK TASK RESPONSES

Chapter 6

Two-dimensional drawing

Packaging nets, task 1 (page 62)
Answer: 2
Packaging nets, task 2 (page 63)
Answer: 1
Packaging nets, task 3 (page 63)
Answer: 1/B, 2/A, 3/A
Orthogonal drawing, task 1 (pages 64–5)
Answer: 1/C, 2/A, 3/A, 4/C
Orthogonal drawing, task 2 (page 65)
Answer: 1/A, 2/C, 3/C, 4A

Third-angle projection symbol, task 3 (page 66)

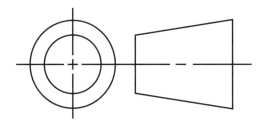

Hidden detail, task 1 (page 67)

1

2

3

4

5

6

ISBN 9780170401791

Hidden detail, task 2 (page 67)

1
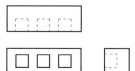

2

3

4

5

6

Dimensioning: centre lines, task 1 (page 68)

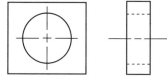

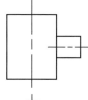

Dimensioning: missing details, task 2 (page 68)

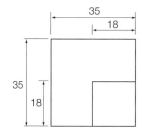

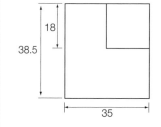

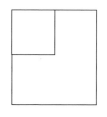

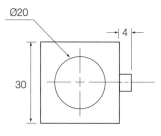

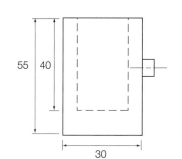

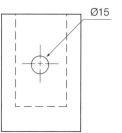

ISBN 9780170401791

Dimensioning, task 3 (page 69)

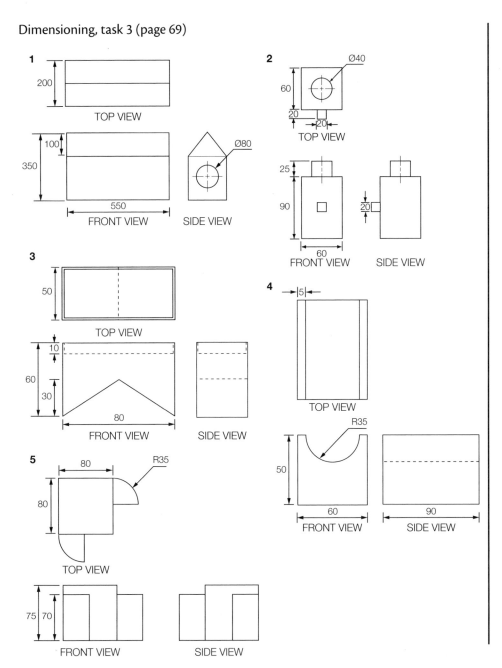

Dimensioning, task 5 (page 70)

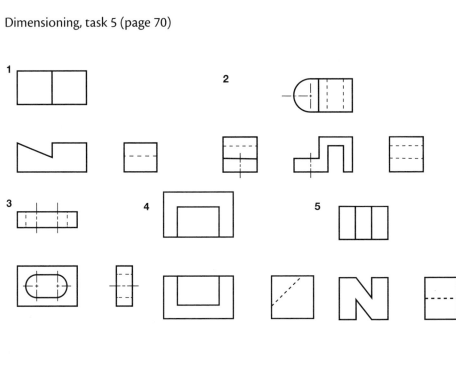

ISBN 9780170401791

Cross-sections, task 1 (page 73)

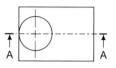

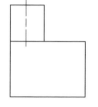 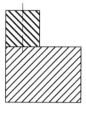 

SECTION A-A

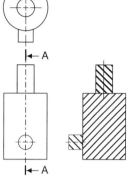

SECTION A-A

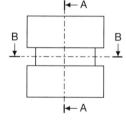 

SECTION A-A   SECTION B-B

Cross-sections, task 2 (page 74)

Answer: 1 = d, 2 = e, 3 = f + b, 4 = c, 5 = a

Cross-sections, task 3 (page 75)

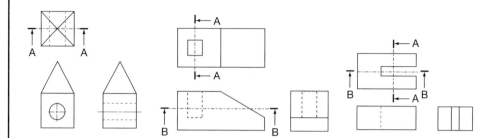

SECTION A-A   SECTION A-A   SECTION B-B   SECTION A-A   SECTION B-B

ISBN 9780170401791

Chapter 7

Three-dimensional drawing

Isometric drawing, task 1 (page 80)

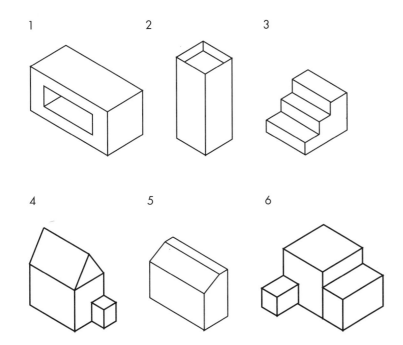

Planometric drawing, task 3 (page 85)

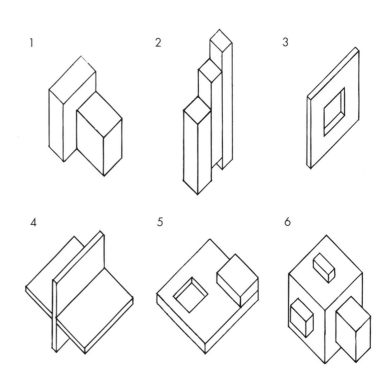

ISBN 9780170401791

Paraline ellipses, task 2 (page 87)

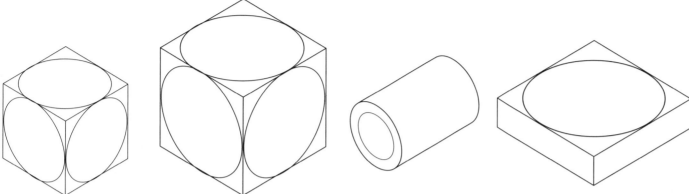

Perspective ellipses, task 3 (page 88)

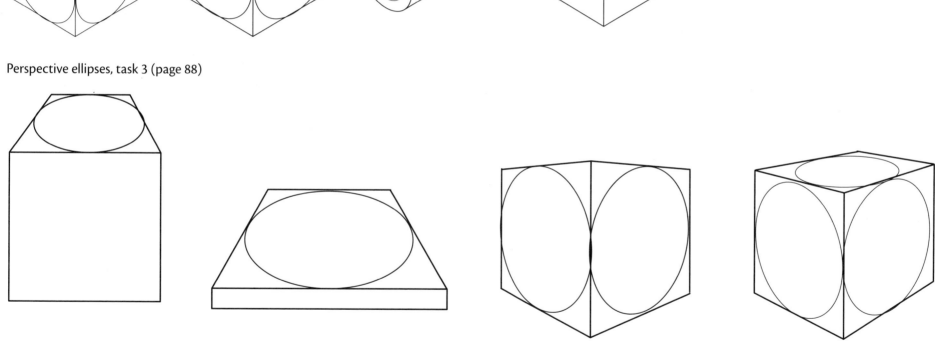

ISBN 9780170401791

One-point perspective drawing, task 3 (page 90)

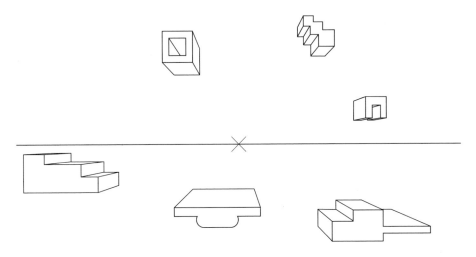

Chapter 9

Typography & layout

The language of typography, task 1 (page 119)

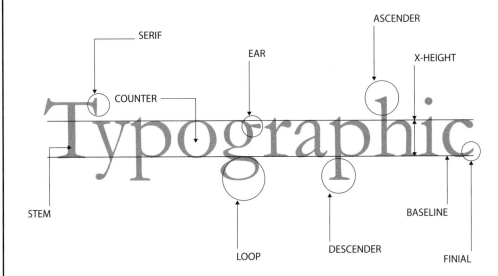

The language of typography, task 2 (page 120)
Answers:
Bracketed serif
Block or Slab serif
Sans serif
Non-bracketed serif

ISBN 9780170401791

NOTES

ISBN 9780170401791

ISBN 9780170401791

APPENDIX III: SELECTED COLOUR IMAGES

Atelier Art Supplies Pty Ltd

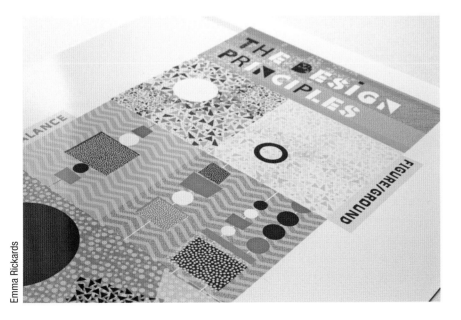

Emma Rickards

iStock.com/Alex_Bond

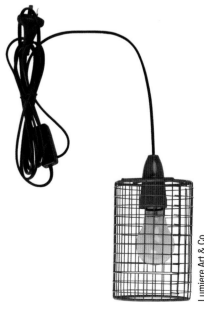

Lumiere Art & Co

ISBN 9780170401791

HALI RUGS. THE INSPIRATION
TO CREATE A BEAUTIFUL ROOM.

Hali Rugs

Ryan Wheatley

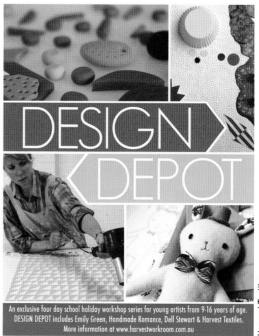

Harvest Textiles

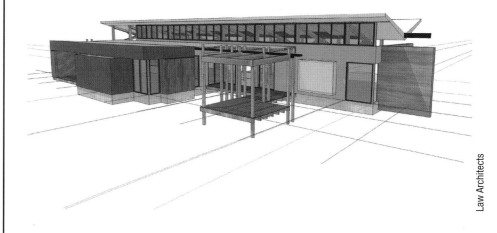

Law Architects

ISBN 9780170401791

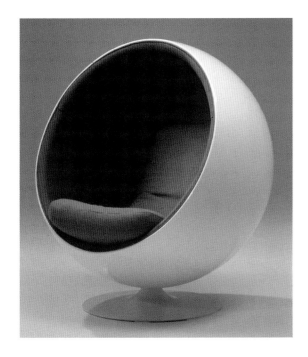

ISBN 9780170401791

iStockphoto.com: YanaZ, anderm, LiliGraphie, Leegudim, GoodLifeStudio. Shutterstock.com: Elena Sistaliuk, gomolach, LanaSweet, Wehands, Alessandro Colle, Everything.

Prue Edmunds

iStock.com/Eva-Katalin

iStock.com/michaeljung

ISBN 9780170401791

Shutterstock.com/Monkey Business Images

iStock.com/Ximagination

iStock.com/max-kegfire

iStock.com/Yuri_Arcurs

ISBN 9780170401791

iStock.com/freestylephoto

Tom Grech

ISBN 9780170401791

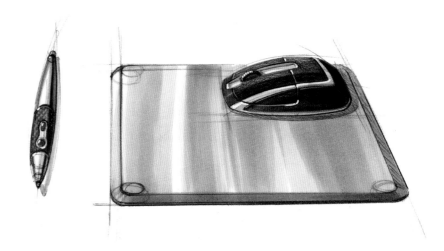

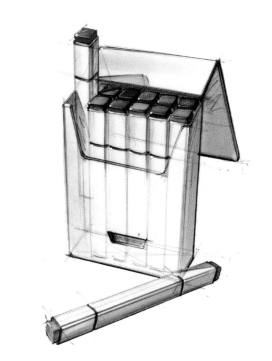

Mark Wilken (all on page)

ISBN 9780170401791

ISBN 9780170401791

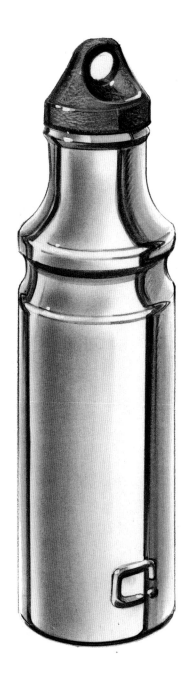

Mark Wilken (all on page)

ISBN 9780170401791

ISBN 9780170401791

ISBN 9780170401791

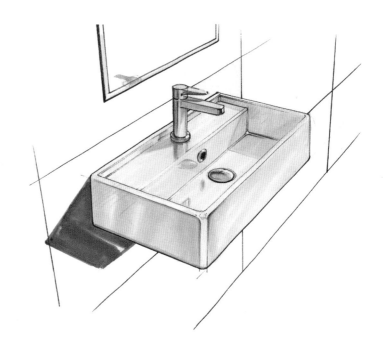

ISBN 9780170401791

Mark Wilken

Mark Wilken

Sophie Coleman

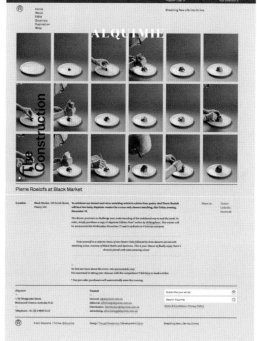

Nicolas Cary

ISBN 9780170401791

Ryan Wheatley (left)

ISBN 9780170401791

CONFIRM 9AM MEETING
EMAIL ALEX
FINALISE SPREADSHEETS
CHECK PRESENTATION
PACK SUITCASE
PASSPORT
LAPTOP
ONLINE CHECK IN
TAKE PHONE OFF CHARGER

Forgotten something important?

You're busy.
Your to-do list is long.
You're late for the plane.
Your phone is still in your hotel room!

Don't panic.

MYCONCIERGE™ service will supply a new phone and deliver yours to your destination within 24 hours.

Business subscriptions from $99 per month*

myconcierge.com.au

We're one step ahead for you

*for approved customers only

STERLING CRUISES

A grand ocean voyage inspired by the glories of travels golden age.

This is the trip of a lifetime and the opportunity to view locations many can only dream of. Every moment on board the magnificent Queen of the Seas is an unforgettable experience. We invite you to become a part of an experience beyond compare.

A Sterling World Cruise.

Reserve your place in history.
Call Sterling Cruises now 13 17 65
or visit sterlingcruises.com

iStockphoto/garywg (ship)

Mark Wilken

Mark Wilken

ISBN 9780170401791

Mark Wilken

Alamy Stock Photo/Mario Mitsis

Shutterstock.com/Savvapanf Photo

ISBN 9780170401791